OURS TO HACK AND TO OWN

OURS TO HACK AND TO OWN

THE RISE OF PLATFORM COOPERATIVISM, A NEW VISION FOR THE FUTURE OF WORK AND A FAIRER INTERNET

EDITED BY
TREBOR SCHOLZ
AND NATHAN SCHNEIDER

O/R

OR Books

New York · London

TABLE OF CONTENTS

PART 1

SOMETHING TO SAY YES TO

1. WHAT THIS IS AND ISN'T ABOUT

TREBOR SCHOLZ AND NATHAN SCHNEIDER

This is a guidebook for a fairer kind of Internet. While we intend to foster something new in the online economy, we do so by turning to something old: the long tradition of cooperative enterprise. The problems of labor abuse and surveillance that have arisen with the "sharing economy," also, are not entirely new; they have much in common with struggles on nineteenth-century factory floors. By considering the emerging platforms in light of well-hewn cooperative principles and practices, we find an optimistic vision for the future of work and life.

Already, this strategy is catching on. Workers, organizers, developers, and social entrepreneurs around the world are experimenting with cooperative platforms and forming conversations about platform cooperativism. This book, therefore, is an effort to serve a movement in the making, to add to the momentum we and our fellow contributors already feel.

We each came to platform cooperativism by somewhat separate paths. Trebor had been convening the Digital Labor conferences at The New School since 2009, from which arose an earlier book, *The Internet as Playground and Factory*. In publications like *The Nation* and *Vice*, Nathan was reporting on the protest movements of 2011 and efforts among young people to create ethical livelihoods, online and off, once the protests receded. We met at OuiShare Fest in Paris in 2014, and, at Trebor's "Sweatshops, Picket Lines, and Barricades" conference later the same year, we both sensed it was time to think about constructive alternatives to the dominant Silicon Valley model.

That December, Trebor published "Platform Cooperativism vs. the Sharing Economy," framing this concept that would come to be this movement's moniker. The same month, *Shareable* published Nathan's article "Owning Is the New Sharing," which mapped out some of the efforts to build cooperative platforms already underway. Realizing our common interest, we discussed these ideas with interested platform-workers, labor advocates, techies, and luddites—many of whom, we found, were venturing into various forms of platform cooperativism already. We agreed it was time that they should meet each other.

In November 2015, we held a two-day event called "Platform Cooperativism: The Internet, Ownership, Democracy" at The New School. More than a thousand people came, including New York City Council members, CEOs, investors, platform creators, and leading scholars. *The Washington Post* deemed the event "a huge success." Shortly after, the Rosa Luxemburg Foundation published Trebor's primer on platform cooperativism, which has been translated into at least seven languages. Follow-up events have taken place in Barcelona, Berlin, Bologna, Boulder, London, Melbourne, Paris, Rome, Milan, Vancouver, and elsewhere.

Before we get started, let's make sure we are talking about the same thing: shared governance and shared ownership of the Internet's levers of power—its platforms and protocols. Democratic ownership and governance are the pillars of what *cooperativism* refers to, both here and historically; without these, the word rings hollow.

Second, this book calls for a process, not another trick of technological solutionism. Platform cooperativism will not come about simply through a few killer apps; it will require a different kind of ecosystem—with appropriate forms of finance, law, policy, and culture—to support the development of democratic online enterprises. This means challenging the cooperative movement to meet the opportunities of the platform economy, and challenging the platform economy to overcome its obsession with short-term profits for the few.

Platform cooperativism is a radical horizon, to be sure, but we should not regard it as an absolute. There will be multiple and sometimes partial means of getting there. A company that shares some ownership and governance is better than one that shares none, and we celebrate that. We encourage a variety of strategies and experiments.

The contributors to this book, in that spirit, represent diverse approaches and perspectives. It is a means of sharing what we learned from the 2015 Platform Cooperativism event more widely, and of drawing more people into the work of overcoming the challenges we face. We can keep the conversation going at platformcoop.net, a place for ongoing discussion of news, resources, and ideas.

After these introductory chapters, the first set of essays considers the opportunities and challenges of the existing online economy, demonstrating the need for more cooperative approaches. The second section addresses the practical design and development of cooperative online platforms; it includes "showcases" of actually existing and in-development platform co-ops. In the third section, we step back to consider the broader ecosystems that we'll need to develop if we are serious about making shared ownership and governance a new norm for the Internet; here, too, are showcases that show how far the platform co-op ecosystem has already come.

Throughout this process, we have been amazed by the enthusiasm and experience that so many people around the world have brought to the #platformcoop conversation, and the effort to make it a reality. We hope this book does justice to the power of what is already underway, as well as the hurdles we still face together. We dedicate the book to those with the courage and imagination to create an Internet worthy of the people it connects.

2. THE MEANINGS OF WORDS

NATHAN SCHNEIDER

For most of the last decade, I've been a reporter, covering stories on how technology is reshaping public life, from debates about God to protests in the streets. One thing I've noticed is that Internet culture has an odd way of using a really important word: *democracy.* When a new app is said to be democratizing something—whether robotic personal assistants or sepia-toned selfies—it means allowing more people to access that something. Just access, along with a big, fat terms of service. Gone are those old associations of town meetings and voting booths; gone are co-ownership, co-governance, and accountability.

Words are the tools of my trade as a writer, so I like to have a handle on what they mean. We rely on them so much. They connect us to each other; they remind us what we're capable of. And I hope that the Internet can help us make our definitions of *democracy* more ambitious, rather than redefining it out of existence.

In late 2014 I was reporting a story about Amazon's Mechanical Turk platform, a website where users can find entirely online piecework—jobs that might take between seconds and hours, like transcribing a receipt, providing feedback on an ad, or taking a sociological survey. I went to Trebor Scholz's Digital Labor conference in New York, which included real-life Mechanical Turkers. One was a wife whose husband lost his job, for instance; another was a former cable technician. I heard them describing what working on the platform is like. Employers can review them, but they can't review employers. Their work can be rejected with no remuneration or recourse. There are no constraints to prevent below-minimum-wage pay. One of them complained in the media and her account was frozen.

Over the course of those days, a kind of question kept coming up among the Turkers, a thought experiment. They wondered aloud: What if *we* owned the platform? How would *we* set the rules?

They'd sit with that for a minute or two, batting ideas back and forth about how to make the platform better for themselves—and for Amazon. Reasonable ideas. Clever ones. But then the ideas would fade back into reality again: back to the complaints.

Since then the agonies over the dictionary-altering Internet have only intensified. People have blockaded Google Buses to protest wealth inequality in San Francisco, and Uber drivers have gone on strike around the world. Increasingly this online economy is becoming *the* economy—the way more and more of us find jobs, relationships, and a roof over our heads. Internet companies aspire to network and monetize everything from our cars to our refrigerators; the companies call this the "Internet of things." But the Turkers' questions have kept coming back to me.

Were they on to something? What if the platforms and networks really were ours? What if we had an Internet of ownership?

REAL SHARING, REAL DEMOCRACY

Another word that the Internet has gotten to is *sharing*. Sharing used to mean something we do with the people we know and trust. In the so-called sharing economy, it means more convenient transactions that take place on distant servers somewhere. Convenience is great, but all along there has been a *real* sharing economy at work, the cooperative economy.

One can trace the modern cooperative movement to the Rochdale Principles of 1844, in England, though it had precursors among ancient tribes, monasteries, and guilds around the world. The rudiments of this stuff could be basic common sense: shared ownership and governance among people who depend on an enterprise, shared profits, and coordination among enterprises rather than competition.

We might not know it, but co-ops are all around us. In Colorado, where I live, 70 percent of the state's territory gets its power from cooperative electric companies that date to the 1930s and earlier, owned and governed by the people they serve. The credit union where I'm a member is one of the top mortgage lenders in the region. Up in the mountains west of me, some years back, a group of neighbors started their own co-op Internet service provider. There's also Land O'Lakes, Organic Valley, and REI.

Co-ops come in all shapes and sizes. They fail less than other businesses, and they often pay better wages (except to top executives). Democracy, it turns out, works—though it can be less lucrative for those just trying to get rich. People in charge are harder to swindle.

I lived in a co-op house once; it followed a certain dirty, organic, folk-music-every-night stereotype. The same couldn't be said, though, for what I saw at Kenya's business school for managers of cooperatives. There, co-ops hold about half the GDP, and those students looked like business students anywhere—except that, along with all the marketing and case studies, they were also learning how to run a company where the people who work for you are your bosses. In the area around Barcelona, among the thousands of members of the Catalan Integral Cooperative, I got a glimpse of what twenty-first-century cooperatives might look like. Rather than securing old-fashioned jobs, these independent workers help each other become less dependent on salaries, and more able to rely on the housing, food, childcare, and computer code they hold in common. They trade with their own digital currency. In cases like this, the traditional lines between workers, producers, consumers, and depositors may become harder to draw.

Part of the cooperative legacy has played out in tech culture already. The Internet relies on free, open-source tools built through feats of peer-to-peer self-governance, like Wikipedia and Linux. Visit many tech offices, from a startup's garage to the Googleplex, and there are self-organizing teams creating projects from the bottom up. Yet somehow this democracy doesn't seem to make it to the boardroom; things are still pretty twentieth-century corporate in there, with whoever happens to own the most shares calling the shots. There's a firewall. We can practice democracy everywhere, it seems, except where it really matters.

There are some pretty sci-fi questions before us these days: Will apps and robots replace our jobs? Will any aspect of our digital lives escape the notice of surveillance? Can there be a digital utopia without the dystopias of sweatshops and blood minerals? In each case the cooperative tradition poses necessary questions, which in the onrush of change we may neglect to ask: Who owns the tools we live by, and how are they governed?

PLATFORM COMMONS

Cooperative enterprises of the past and present have relied on two kinds of strategies to gain a foothold in economies and cultures premised on competition. One is the competitive advantage to be found in cooperation—the ability to succeed where conventional markets fail, for instance, and the power latent in solidarity. The second is when the rules of the system are changed to support more cooperative practices—especially through governments that see the value of cooperative enterprise enough to encourage and fund it. For platform cooperativism to flourish, I suspect we need both of these.

We can begin by identifying the competitive advantages of cooperation. Cooperative practices, for instance, are poised to thicken the notoriously loose ties that online connectedness normally offers. And as big tech companies continue having difficulty treating workers and users as—well, people—co-ops can offer positive, ethical alternatives that workers and users can turn to. Hybrid models—combining aspects of a conventional company with aspects of cooperative ownership and governance—seem promising in the short term. Yet the rules of the system remain very much tilted against cooperativism.

17

This needs to change. Governments should recognize that cooperative platforms will mean more wealth staying in their communities and serving their constituents. Rather than trying (and failing) to say "no" to the likes of Uber, platform co-ops are something public institutions can say "yes" to. We need laws that make it easier to form and finance co-ops, as well as public investment in business development—stuff that extractive businesses get all the time.

This also means thinking differently about the incumbents. The Facebooks, Googles, and Ubers aren't just regular companies anymore. Their business models are based on how dependent so many of us are on them; their ubiquity, in turn, is what makes them useful. They're becoming public utilities. The less we have a choice about whether to use them, the more we need democracy to step in. What if a new generation of antitrust laws, instead of breaking up the emerging online utilities, created a pathway to more democratic ownership?

Rather than donating Facebook shares to his own LLC, Mark Zuckerberg could put them into a trust owned and controlled by Facebook users themselves. Then they, too, could have a seat in the boardroom when decisions are made about what to do with all that valuable personal data they pour into the platform—and they'd have a stake in ensuring the platform succeeds. How would you vote?

These aren't just questions about what kind of Internet we want, or even what kind of world we want; they're about how we see ourselves. Do we trust ourselves enough to expect democracy from the institutions on which we rely? Are we bold enough to imagine, as the Mechanical Turkers were, what the Internet would look like if we were in charge?

Thirty years ago, when the Internet wasn't much more than a lab experiment, the social critic Theodore Roszak saw a lot of this coming. "Making the democratic most of the Information Age," he wrote in *The Cult of Information*, "is a matter not only of technology but also of the social organization of that technology."

We forget that. New gizmos come and go so quickly that we hardly notice when the meanings of our words change, and when what we expect of ourselves changes with them. Ordinary people have already made the Internet their own with their hacks, their memes, their protests, and their dreams. The cost of forfeiting control over these things is too high, and too mysterious. We need to expect better, to demand more. It's time that we own and govern what is ours already.

3. HOW PLATFORM COOPERATIVISM CAN UNLEASH THE NETWORK

TREBOR SCHOLZ

In 1998 I moved into a small Buddhist temple in San Francisco's Mission District. My spiritual comrades in this commune could not understand why I would spend all the money that I had saved on an IBM laptop when the community already owned a computer. As someone who studies the social impact of the Internet, I was surprised by the proposal to collectively use one computer. For me, up to that point, thinking about the Internet meant thinking about individual use, not communal ownership. This episode showed me how a culture of genuine sharing can also mean sharing technology, just like anything else.

Over the past five years, the technological ingenuity of the "sharing economy" deeply resonated with the zeitgeist. Emphasizing community, underutilized resources, and open data, the genuine sharing economy was initially presented as a challenge to corporate power. Just like my Buddhist friends, the pioneers of this economy proposed to split the use of lawn mowers, drills, and cars. But soon, the non-commercial values behind many platforms were rewritten in the boardrooms of Silicon Valley, turning the "sharing economy" into a misnomer. Today, facing various prophecies about sharing and the future of work, we need to remind ourselves that there is no unstoppable evolution leading to the uberization of society; more positive alternatives are possible.

In *Average Is Over*, the economist Tyler Cowen foresees a future in which a tiny "hyper meritocracy" would make millions while the rest of us struggle to survive on anywhere between $5,000 and $10,000 a year. It already works quite well in Mexico, Cowen quips. Carl B. Frey and Michael A. Osborne predict that 47 percent of all jobs are at risk of being automated over the next twenty years. And I have no doubt about the vision of platform owners like Travis Kalanick (Uber), Jeff Bezos (Amazon), or Lukas Biewald (CrowdFlower)—who, in the absence of government regulation and resistance from workers, will simply exploit their undervalued workers. I'm all on board for Paul Mason's and Kathi Weeks' visions for a post-capitalist, post-work future where universal basic income will rule the way we think about life opportunities. In the United States, however, unlike in Finland, the chances for this scenario becoming a reality over the next two years are not high. The question then becomes what we can do right now, with and for the most precarious among the contingent third of the American workforce, which is unlikely to see the return of the traditional safety net, the forty-hour workweek, or a steady paycheck.

Today's Internet bears little resemblance to the ARPA-designed, non-commercial, decentralized, post-Sputnik network. We are finding that the sources of our entertainment, the platforms where we are logging on to work every day, and the apps that constantly draw us into feedback loops are all owned by a small number of deep-pocketed founders and stockholders. That's simply unacceptable, and it is for this reason that I proposed a theory of "platform cooperativism" in 2014. Workers in the on-demand economy are called upon to "live like lions," but with slightly more flexibility have come more risks and harsher taskmasters. The average on-demand economy worker earns $7,900 a year through labor platforms, which indicates that many of them work only part-time in this digital economy. Often disregarded in this discussion are those who are pushed out of the market by, for example, Uber drivers, who are 40 percent college-educated and more likely to be white than legacy taxi drivers who may lose their jobs.

Many of the business models of the "sharing economy" are based on the strategic nullification of the law. Companies knowingly violate city regulations and labor laws. This allows them to undermine the competition and then point to a large customer base to demand legislative changes that benefit their dubious modus operandi. Firms are also activating their app-based consumers as a grassroots political movement to help them lobby for corporate interests. Privacy should be a concern for workers and customers, too. Uber is analyzing the routines of its customers, from their commutes to their one-night stands, to then impose surge pricing when they most rely on the service. Navigating legal gray zones, these deregulated commerce hubs sometimes misclassify employees as independent contractors. They are labeling them "turkers," "driver-partners," or "rabbits," but never workers. Hiding behind the curtain of the Internet, they would like us to believe that they are tech rather than labor companies.

In the decade between 2000 and 2010, the median income in the United States declined by 7 percent when adjusted for inflation. In 2014 51 percent of Americans made less than $30,000 a year, and 76 percent of them had no savings whatsoever. Since the 1970s, we have witnessed concerted efforts to move people out of direct employment, which has led to the steady growth of the number of independent contractors and freelancers. Digital labor, a child of the low-wage crisis, is part of that process.

What has the "sharing economy" really gotten us? Beyond the consumer convenience and efficiency in creating short-term profits for the few, it has demonstrated how, in terms of social well-being and environmental sustainability, capitalism turns out to be amazingly ineffective in watching out for people. Seemingly overnight, the gains of more than one hundred years of labor struggles, dating back to the Haymarket Riots in 1886 and the protests after the Shirtwaist Factory fire in 1911, have been stalled. Also, the Fair Labor Standards Act of 1938 suddenly has far less pull because the number of employees is shrinking rapidly.

Among all the problems of the twenty-first century that are related to workers—inequality, stagnant wages, loss of rights—the biggest predicament is that there seem to be so few realistic alternatives. But there are. I will identify four approaches.

The first two approaches are based on the belief in negotiation with corporate owners and with government. The Domestic Workers Alliance, for example, formulated a Good Work Code in hopes that policy makers would endorse their guidelines and that platform owners would follow them. Seattle imposed a tax on Uber and gave drivers the right to unionize, Mayor Bill de Blasio of New York City made attempts to curb the number of Uber cars, and the city of San Francisco tried to regulate Airbnb. A third pathway is to move production outside of the market altogether. Yochai Benkler labeled this "non-market peer production," with the most successful example being Wikipedia. And, finally, for the compensated labor market, there is a fourth approach, which is platform cooperativism, a model of social organization based on the understanding that it is hard to substantially change what you don't own.

My thinking about platform cooperativism owes much to the Digital Labor conferences at The New School. These events started in 2009 and one of the recent ones was Platform Cooperativism in 2015. Initially, at these events, discussions focused on the Italian Workerists, immaterial labor, and "playbor." Artists like Burak Arikan, Alex Rivera, Stephanie Rothenberg, and Dmytri Kleiner played pioneering roles in alerting the public to these issues. Later, debates became more concerned with "crowd fleecing," the exploitation of thousands of invisible workers in crowdsourcing systems like Amazon Mechanical Turk or content moderation farms in the Philippines. Over the past few years, the search for concrete alternatives for a better future of work has become more dynamic.

The theory of platform cooperativism has two main tenets: communal ownership and democratic governance. It is bringing together 135 years of worker self-management, the roughly 170 years of the cooperative movement, and commons-based peer production with the compensated digital economy. The term "platform" refers to places where we hang out, work, tinker, and generate value after we switch on our phones or computers. The "cooperativism" part is about an ownership model for labor and logistics platforms or online marketplaces that replaces the likes of Uber with cooperatives, communities, cities, or inventive unions. These new structures embrace the technology to creatively reshape it, embed their values, and then operate it in support of local economies. Seriously, why does a village in Denmark or a town like Marfa in rural West Texas have to generate profits for some fifty people in Silicon Valley if they can create their own version of Airbnb? Instead of trying to be the next Silicon Valley, generating profits for the few, these cities could mandate the use of a cooperative platform, which could maximize use value for the community.

Platform co-ops already exist, from cooperatively owned online labor brokerages and marketplaces like Fairmondo, to video streaming sites that are owned by filmmakers and their fans. Photographers co-own the stock photography cooperative Stocksy and massage therapists in San Francisco started the freelancer-owned online labor market Loconomics. Students at Cornell University built Coopify for (and with) co-ops of low-income immigrants in Sunset Park, Brooklyn. Platform co-ops could be attractive options for home health care professionals and also low-income residents, or pensioners who need to earn extra cash. In the United States, the 650,000 people who are released from U.S. prisons every year would be likely to welcome dignified work. And finally, platform co-ops might be attractive for refugees, for whom it often takes as long as eight years after their immigration to find a job, even in a country like Sweden. With this model, workers can become collective owners; they do no longer have to subscribe to the pathology of the old system that trained them to be followers.

Few people will feel drawn to build a platform co-op based on abstract principles. But for the already committed, common principles and values matter. From the Rochdale Society of Equitable Pioneers, to African-American cooperatives in the South of the United States, to the Mondragon Corporation in Spain, forming any kind of cooperative endeavor has always started with a study group. Political scientist Elinor Ostrom reminded us that aspiring to create alternatives without rigorous study is a pipe dream, a vain hope. Being realistic about cooperative culture is essential. From the history of cooperatives in the United States, we learned that they are indeed able to offer a more stable income and a dignified workplace. While the necessary enthusiasm of makers doesn't always sit well with justifiably skeptical scholars, their dialogue is important. Jointly, they could rewrite the Rochdale principles for the digital economy, for instance. Education is an essential cornerstone of platform cooperativism.

Platform co-ops should consider the following principles. The first one, which I explained already, is communal ownership of platforms and protocols. Second, platform co-ops have to be able to offer income security and good pay for all people working for the co-op. And history shows that co-ops are able to offer this. Emilia-Romagna, an area in Italy that encouraged employee ownership, consumer cooperatives, and agricultural co-ops, has lower unemployment than other regions in Italy. The flagship of cooperatives, Mondragon, is a network of co-ops that employed 74,061 people in 2013. But in the United States, despite its dominance in areas like orange juice production, the cooperative model has been faced with many challenges, including competition with multinational corporate giants, public awareness, self-exploitation, and the network effect. So, it is essential for platform co-ops to study the communities they'd like to serve and get their value proposition right.

In opposition to the black-box systems of the Snowden-era Internet, these platforms need to distinguish themselves by making their data flows transparent. They need to show where the data about customers and workers are stored, to whom they are sold, and for what purpose. Work on platform co-ops needs to be co-determined. The people who are meant to populate the platform in the end must be involved in its design from the very beginning. They need to understand the parameters and patterns that govern their working environment. A protective legal framework is not only essential to guarantee the right to organize and the freedom of expression but it can help to guard against platform-based child labor, wage theft, arbitrary behavior, litigation, and excessive workplace surveillance along the lines of the "reputation systems" of companies like Lyft and Uber that "deactivate" drivers if their ratings fall below 4.5 stars. Crowd workers should have a right to know what they are working on instead of contributing to mysterious projects posted by anonymous consignors.

At its heart, platform cooperativism is not about any particular technology but the politics of lived acts of cooperation. Soon, we may no longer have to contend with websites and apps but, more and more, with 5G wireless services (more mobile work), protocols, and AI. We have to design for tomorrow's labor market. In the absence of rigorous democratic debates, online labor behemoths are producing their version of the future of work right in front of us. We have to move quickly. Together with cities like Berlin, Barcelona, Paris, and Rio de Janeiro, which have already pushed back against Uber and Airbnb, we ought to refine the discourse around "smart cities" and machine ownership. We need incubators, small experiments, step-by-step walkthroughs, best practices, and legal templates that online co-ops can use. Developers will script a WordPress for platform co-ops, a free-software labor platform that local developers can customize. Ultimately, platform cooperativism is not merely about countering destructive visions of the future, it is about the marriage of technology and cooperativism and what it can do for our children, our children's children, and their children into the future.

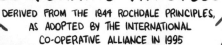

THE SEVEN COOPERATIVE PRINCIPLES

DERIVED FROM THE 1844 ROCHDALE PRINCIPLES,
AS ADOPTED BY THE INTERNATIONAL
CO-OPERATIVE ALLIANCE IN 1995

BY
SUSIE CAGLE

← → ⟳ ⌂ | https://platformnationcoop.org

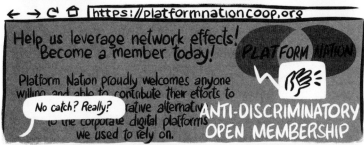

Help us leverage network effects!
Become a member today!

PLATFORM NATION

Platform Nation proudly welcomes anyone
willing and able to contribute their efforts to
_____ rative alternative
to the corporate digital platforms
we used to rely on.

No catch? Really?

ANTI-DISCRIMINATORY OPEN MEMBERSHIP

RANK YOUR CHOICES:

DEMOCRATIC MEMBER CONTROL

EQUITABLE MEMBER ECONOMIC PARTICIPATION

TIMEBANKR
CONVERT YOUR INVESTMENTS
BETWEEN CASH AND LABOR
WITH FULL TRANSPARENCY!

CONTINUE   NO, I HATE SHARING

5. EIGHT FACTS ABOUT COOPERATIVE ENTERPRISE

JESSICA GORDON NEMBHARD

1. **Cooperative enterprises address market failure and need.** They provide rural electricity or other utilities in sparsely populated areas; affordable healthy and organic foods, especially in food deserts; access to credit and banking services; access to affordable housing; access to quality affordable child or elder care; and access to markets for culturally sensitive goods and arts.

2. **Cooperatives overcome historical barriers to development** in the ways they aggregate people, resources, and capital. Of 162 non-agricultural cooperatives in one study, 44 percent of the respondents said they could not have opened their business had it not been organized as a cooperative.

3. **The economic activity of the approximately thirty thousand cooperatives in the United States contributes an estimated $154 billion to the nation's total income.** Co-ops have helped to create over 2.1 million jobs, with an impact on wages and salaries of almost $75 billion. After becoming owners of a house-cleaning co-op in Oakland, the workers experienced a median income increase from $24,000 to over $40,000.

4. **Cooperative businesses have lower failure rates than other businesses,** both after the first year (10 percent failure versus 60-80 percent) and after five years (90 percent still operating versus 3-5 percent). Evidence also shows that cooperatives successfully address the effects of economic crises and survive crises better.

5. **Cooperatives are more likely to promote community growth** than an investor-oriented firm, since most are owned and controlled by local residents. Since cooperative business objectives are needs-oriented, cooperatives are more likely to stay in the communities where they originate. For every $1,000 spent at a food co-op, $1,606 goes to the local economy; for every $1 million in sales, 9.3 jobs are created.

6. **Cooperative businesses stabilize communities** because they serve as business anchors, distributing, recycling, and multiplying local expertise and capital. They enable their owners to generate income and jobs; accumulate assets; provide affordable, quality goods and services; and develop human and social capital.

7. **Co-ops and their members pay taxes and are good citizens.** They tend to give donations to their communities, pay their employees fairly, and use sustainable business practices.

8. **Cooperative start-up costs can be low.** Members can contribute time and capital, offsetting costs that require other businesses to seek outside financing. Co-ops are also eligible to apply for loans and grants from a number of federal and state agencies designed to support co-op development, and are often provided relatively low-cost loans from non-governmental financial institutions like cooperative banks.

Adapted from Benefits and Impacts of Cooperatives, *working white paper for the Center on Race and Wealth, Howard University (February 2014),* http://is.gd/ItoPHT.

PART 2

PLATFORM CAPITALISM

6. RENAISSANCE NOW

DOUGLAS RUSHKOFF

What would it take to make the digital economy less like industrial capitalism on steroids, and more consonant with the distributive nature of digital networks themselves? People are trying a lot of strategies, from peer-to-peer value exchanges and the restoration of the commons to crowd-funded debt remediation schemes and local favor-banks. Something big is going on here.

Surprisingly, perhaps, these efforts rarely involve digital technology itself at their core. Rather, they are informed by a digital sensibility. It turns out that we don't actually need blockchains to reconcile and administrate the contributions that each driver has made to a driver-owned version of Uber any more than we need cryptography to stage a debt strike.

Such activities are not so much digital in their implementation as they are in the hands-on, hacker ethos from which they emerge. *Digital*, after all, refers first and foremost to the fingers—the digits—through which human beings create value. In a sense, the digital hearkens back in time, not just ahead, to a time when people were not disconnected from the value they created, and when the world was not simply a set of resources to be extracted by corporations.

One wouldn't know that from looking at the dominant players in the digital economy today. Instead of remaking the economy from the ground up, these companies—Amazon, Uber, Facebook, Apple...take your pick—simply practice capitalism with digital tools. Their founders are happy to "disrupt" one industry or another, but they never even consider disrupting the landscape on which they are functioning—the operating system of venture capital beneath all the apps and devices they make.

As soon as a developer comes up with a potentially useful digital technology, he (yes, usually a he) runs to a venture capitalist or investment banker for funds. Those funders then run the show. Satisfied with nothing less than 100x returns on their money, they push the founders to "pivot" the business toward outlandish, "home run" outcomes. The object of the game is not to create a successful business, but to "exit" through an IPO or acquisition before the business fails. In spite of their abuse of the environmentalist's lexicon, they do not create sustainable "ecosystems" at all, but scorched-earth monopolies through which no one—no one—gets to create or exchange value.

That doesn't really matter. All they have to do is extract enough value from people and places in order to sell themselves to someone else—or leverage their monopoly in one market (like books or taxis) to another one (like movies or robotic vehicles).

Looked at from a digital perspective, these companies are really just software, optimized to extract as much value as they can from the real world, and convert it into share price for their investors. They take real, working, circulating currency, and turn it into frozen, static, useless capital. That's the digitally enabled division of wealth, in a nutshell. It's not truly digital; it's not hands-on, connective, or a hack of the underlying operating system. It's the same old industrialism, being practiced with powerful new digital tools. It's also utterly inconsistent with the underlying biases of digital technology. That's why such schemes tend to work against the interests of real people or communities, and are bound to fail in the long run.

Industrialism, an outcome of the Renaissance, worked pretty well as long as the economy was growing. Based on the premise of debt-based central currency—interest-bearing bank notes—the object of industrialism was to grow the economy so that more money could be paid back to lenders than was borrowed.

Industrialism replaced the peer-to-peer economy of the late Middle Ages. Skilled workers were shunned in favor of low-cost assembly line laborers. Local currencies that promoted frictionless exchange were replaced by high-cost, interest-bearing money. Human connections between interdependent producers and service providers were overtaken by artificial connections between brands and consumers. Acquisition became a human value more important than pleasure itself, as we all (with a bit of help from marketing psychologists) took on the characteristics of competitive businesses in our daily lives and interactions. America became a Tupperware party.

All this was engineered simply to extract value from the periphery to the middle—from the real and ground-up to the abstract stocks and bonds of the already wealthy. All in accordance with the underlying biases of the Renaissance: the centralization of power, the rise of the individual, the emergence of the chartered monopoly, and the spread of empires to new continents. People and places were just slaves and territories.

The dominant digital economy—the one driven by venture capital, the stock market, and business as usual—expresses these values and exacerbates all the same mechanisms, treating people and places the very same way that Renaissance princes did. Algorithms exacerbate the extractive nature of our markets, while companies like Uber and Airbnb leverage monopolies to disempower labor and neighborhoods. Where territorial expansion once supplied corporations with new room for growth, in the digital age the only new surface area is human time, awareness, and data. People spend an increasing amount of their lives in service of a digital economy that delivers them nothing in return. Meanwhile, the best minds out of MIT and Stanford are hired to optimize every device, app, and operating system to do this more completely.

Of course, those left jobless (or simply incapable of generating income through the same work they've always done for pay) are among the first to challenge the underlying assumptions of the first, faux digital economy. To them—to us—the digital age is still the harbinger of something different than business as usual. It offers not a mere amplification of the worst of capitalism, but the possibility for a state change:

something as different from corporate industrialism as corporate industrialism was from the artisanal marketplace of the late Middle Ages.

Indeed, just as the Renaissance retrieved the values of empire from ancient Rome, and re-birthed them as capitalism and industrialism, might the digital era constitute a renaissance of its own? And if it does, what values will it retrieve?

Well, if history is any lesson, today's renaissance will retrieve the values that were submerged and repressed by the last one. The Renaissance obsolesced medieval attention to craft, quality, and personal connections among participants in the marketplace. Not only were peer-to-peer currencies outlawed, but guilds were disbanded, the commons were privatized, and craftspeople used to being paid for the value they created became wage laborers working by the hour, with no ownership stake in their enterprises.

Laugh all you like at the rise of artisanal beers and hand-knitted sweaters, but these seemingly precious throwbacks augur the retrieval of the medieval sensibility as surely as Burning Man, *Game of Thrones*, and the newly expanded menu of body modifications offered by the piercing place at the mall. We are already retrieving the lost spirit of medievalism in our culture and media.

The migration of this sensibility to our economy is next. And necessary. Through the establishment of guilds, such as the Institute of Electrical and Electronics Engineers, technologists are setting their own standards for how they'll apply their skills—and the price of the NASDAQ is not on their list. Etsy retrieves the spirit of the peer-to-peer marketplace, while the Creative Commons begins to compensate for the privatization of shared intellectual resources.

Online favor-banks, time dollars, and local currencies retrieve the possibility for direct, peer-to-peer exchange of value, while the blockchain obsolesces the monopoly of central authorities over accountability and authentication.

Platform cooperatives—as a direct affront to the platform monopolies characterizing digital industrialism—offer a means of both reclaiming the value we create and forging the solidarity we need to work toward our collective good. Instead of extracting value and delivering it up to distant shareholders, we harvest, circulate, and recycle the value again and again. And those are precisely the habits we must retrieve as we move ahead from an extractive and growth-based economy to one as regenerative and sustainable as we're going to need to survive the great challenges of our time.

As the essays in this book make clear, the renaissance is on. Digital is not simply a new high-tech tool to promote the agenda of the last renaissance. It's a good old human sensibility for bringing on the next one.

7. OLD EXCLUSION IN EMERGENT SPACES

JULIET B. SCHOR

It has been widely assumed by those drawn to the idea of platform cooperativism that such platforms would help enhance gender and race diversity and reduce the inequality that has often prevailed in the online economy so far. But what evidence do we have that this would be the case?

For the last five years, along with a team of PhD graduate students in sociology, I have been studying platforms. We have been generously funded by the MacArthur Foundation, which is interested in how digital technologies are affecting social life, opportunity, participation, and social inequality. The research has mainly been qualitative: based on interviews with people who engage in online platforms, or what we have called the "connected economy," as well as intensive ethnographic research at connected economy sites. We started by concentrating on nonprofit grassroots initiatives that were trying to re-shape the way Americans get access to, exchange, and consume goods and services, such as a time bank and a food swap. We later added educational and learning sites, such as online courses and workshops that offer "upskilling," and a makerspace. Then we moved on to the far more controversial for-profits, such as Airbnb, and peer-to-peer car rental sites. We've also been studying on-demand labor sites, including TaskRabbit, Postmates, and Favor.

We're interested in many aspects of these innovative arrangements. How are they organized? Who is benefiting from them? How do people (on both sides of the markets) feel about them? What are the dynamics of inequality and access that operate in these spaces?

Participants in the platform cooperativism movement have high ambitions to create platforms that are owned and governed by their users, that embody principles of equality and access, and that serve the common good. They hope to develop "community" among participants and to engage people in the maintenance of this community. These are aspirations that are also shared by many of the innovators and members of a number of the nonprofit sites that we have studied. There may be some shared challenges as well.

Consider three of our sites: the time bank, the food swap, and the makerspace. The time bank is a volunteer-run and -led organization that allows people to trade labor services on a purely egalitarian basis. All types of services are welcome, so long as they are legal, and the time bank has a wide diversity of offerings. They are all valued in terms of the time it takes to provide them, irrespective of the market value of the service; an hour of a lawyer or a plumber's time is valued at the same rate as that of a dog walker or someone offering a ride to the airport. Similarly, in the food swap, people prepare foods in bulk and bring them to trade with others' preparations. There is no cost to participate. Here, also, everyone's food is valued more or less equally, whatever its cost in terms of materials or labor; foods can be exchanged at a one-to-one ratio for a roughly equivalent size. Our third site, the makerspace, is also a place that attempts to create an open, accessible site for people to learn to use a group of collectively owned tools, and to become creative "makers." The site offers classes and membership, as well as a lively community of highly engaged individuals.

One of the central questions of our research is to assess the extent to which the aspirations and ambitions of these sites are being realized. Does an egalitarian trading economy work? Can it be expanded to cover more services, more people, and more consumption needs? Does the makerspace succeed at teaching skills and expanding its community?

While there are many successes among our three sites, our research also led to a troubling finding: all three cases are plagued with status-seeking, subtle forms of social exclusion, and non-egalitarian behavior that threatened the core goals of founders and members. They are also all highly racialized sites in which nearly all participants are white. And they are characterized by gender skews as well. The time bank and food swap are strong-majority female. The makerspace is about two-thirds male in terms of participants, but far more male in terms of power, influence, and the distribution of social status. Race, class, and gender inequalities are pervasive in these sites, at times even threatening their viability. Finally—and this is key to many of the dynamics—with the exception of a few makers who gained their expertise outside the formal education system, the people we interviewed and studied have disproportionately high education and high parental education levels. This is what Pierre Bourdieu has called "high cultural capital."

Let's start with the time bank. The time bank has a very white, very female membership whose education levels, as well as parental education levels, are "off the charts" in comparison to the U.S. population. One result is that, despite their ideological alignment with the goals of the time bank, many members subtly undermine its functioning. They do this by believing that their own time is worth more than that of others, by failing to offer highly valued skills that they have (such as computer programming) because they'd rather "work with their hands," or because they screen out potential trading partners on account of grammatical errors or less-than-professional presentation on the time bank's website. In our research, we found that people were often unaware of the ways in which their preferences were undermining the egalitarian goals of the bank. Yet that behavior led to a volume of trading being far below its potential on account of social exclusion.

Low trading volume also proved to be a problem for the food swap. Here, social exclusion took a fairly familiar form: snobbishness around food. Founders and longer-term members were reluctant to trade with newcomers whose offerings did not conform to a strong, albeit unarticulated, ideal. The wrong packaging was often fatal to one's chances of finding trading partners. So was offering the wrong *type* of foods. How one presented the food, the size of the offerings, and the choice of ingredients were critical to making trades. Including any processed ingredients, for example, no matter how artfully combined or "re-mixed," was taboo. In our research, near-moral condemnation of offerings that did not conform resulted in a collapse of the food swap. Newcomers failed to stay as their offerings were rejected. Even longer-term members eventually stopped coming. Socially exclusionary practices among a homogeneous, highly educated group led to the failure of this once-promising social innovation.

The makerspace presented a different kind of social dynamic. Unlike the other two sites, it did have a vibrant community with high demand for services. Many people took classes or joined as members. Space was at a premium. On the surface, the site appeared to be highly successful. But as we did our research, we discovered that it was dominated by a small group of men who vied for status via a strongly classed set of values. At the core of those values was what Bourdieu has called "distance from [economic] necessity." To gain high status at the makerspace, one had to participate in a status system that denigrated usefulness and functionality and valued esoterica, eccentricity, and a certain kind of waste. Makers got status by creating items that would be destroyed after a single use. They specialized in highly abstruse sub-cultural knowledge. They created an exclusive community, which screened out people whose making was ordinary, mundane, or economically useful. In that way, quite unintentionally, they created barriers for ordinary people. As a result, the site was culturally and racially very homogeneous.

In our research on for-profit sharing platforms, we found a more open trading landscape—it was easier for people to join and make exchanges. However, we did also find evidence of discrimination. Airbnb hosts screened potential guests for educational credentials, evidence of financial assets, and the like. Some indicated they only wanted to rent to people like themselves. On the other hand, dynamics of social exclusion were stronger and more prevalent in the nonprofit spaces.

So what's the lesson of our research for platform cooperativism? Historically, cooperatives have mostly been formed by working class people who share a lot in terms of culture and history. That shared culture was crucial to their success. But platform cooperativism is coming from a different social space. If platform co-ops are to succeed without reproducing their own more privileged class, race, and gender homogeneousness, founders and early participants must be highly attuned to subtle social dynamics that valorize the practices and traits of dominant groups. Furthermore, they must stop those dynamics from developing. Practically speaking, achieving that probably means starting with a diverse group of founders and early participants—at the very least on the social dimensions of class, race, and gender.

8. WORSE THAN CAPITALISM

MCKENZIE WARK

What if this was no longer capitalism, but something worse? Such a perspective might help explain some of the features of the contemporary political-economic landscape. My argument, odd though it may sound, is that both capital and labor have lost ground to an emerging ruling class, one that confronts a quite different kind of antagonist.

It helps to see capitalism as already a kind of second-order mode of commodified production. First-order commodification emerged—in part, at least—through the transformation of the relations between peasants and their lords; the peasants lost traditional rights to arable land and to the commons. In place of the (supposedly) ancient rights and duties that held between landlord and peasant, in which the peasant's duties to the lord were paid directly with a share of the produce, the peasant had to pay rent in cash.

First-order commodification was thus the commodification of land. Pieces of land became abstract pieces of property that could be bought and sold. Peasants lost traditional rights to land and saw much of it enclosed and privatized. A "surplus population" of peasants ended up in the cities, where they were to become the working class, sellers of labor-power.

Capitalism was a second-order mode of commodified production, built on top of the pastoral one that preceded it in the countryside. One can forget that when David Ricardo wrote *On the Principles of Political Economy and Taxation,* he wrote on behalf of a rising, urban, capitalist ruling class and against the interests of a pre-existing, rural, pastoralist ruling class. It was a study in *intra*-ruling class studies.

In opposition to the pastoralist ruling class, the capitalist ruling class constructed a rather more abstract mode of production, one in which not only land but labor and the factory could be elaborate forms of fungible private property. With the destruction of the privileges awarded it by the state, the landlord class became a subordinate ruling class within capitalism, still extracting its extortionate ground rents (as indeed it still does today) but unable to claim the whole of the state as its own and to govern exclusively in its own interests.

The peasantry were, of course, no mere spectators upon their own oppression, but resisted the landlord class, every so often rising up against it. However, the peasantry tended toward a politics based on ancient rights. The rise of the modern labor movement was a cultural revolution that replaced the backward-looking peasant politics with a forward-looking one, based on the evident fact of capitalism as the dominant mode of production.

Such might be a more or less orthodox thumbnail sketch of the rise of capitalism in Britain, where it first arose. Of course, elsewhere in the world it followed different paths. But rather than turn toward the complicated business of pluralizing this historical sketch, I want to do something different: to pose the question of whether there is, on top of the second-order commodified mode of production of capitalism, a third-order commodified mode of production—what I will call vectoralism.

First-order commodification, what I call pastoralism, made land into a form of abstract private property relation. Second-order commodification, generally called capitalism, much advanced the abstraction of the private property relation into fungible things. Third-order commodication, which I call vectoralism, extends abstraction much further, subordinating information to whole new kinds of private property rights, and in the process creating new kinds of class relations.

On top of the class relation of landlords and peasants, and of capitalists and the working class, there is a relation between a vectoralist class that owns the vector of information in one form or another, and a hacker class that has to produce new forms of information that can be made into private property.

This emerging class relation does not replace previous layers of commodified abstraction, but it does transform them. Initially, the vectoralist class enabled capital to outwit the working class in the class conflicts of the late twentieth century. The information vector was what enabled capital to route around the power of labor to interrupt production. The information vector enabled capital to draw resources from a variety of sources at short notice. The information vector enabled capital to develop productive resources remote from traditional working class communities, with their historic memory and capacity for self-organization.

In the short term, the vectoralist class was helpful to capital in its struggle against labor, but in the long run, it is trying to subordinate capital to itself. Take a look at the top Fortune 500 companies, or the top "unicorn" venture capital darlings of the moment. With a few exceptions, one finds iterations of the same thing: companies whose power and wealth relies on stocks or flows of information, which control either the extensive vector over space or the intensive vector of an archive of commodified information—so-called intellectual property.

Whether it is finance, tech, cars, drugs, food, or chemicals—often the big companies no longer actually make their products. That can be contracted out to a competing mass of capitalist suppliers. What the vectoralist firm owns and controls is brands, patents, copyrights, and trademarks, or it controls the networks, clouds, and infrastructures, along which such information might move.

The rise of the so-called sharing economy is really just a logical extension of this contracting out of actual material services and labor by firms that control unequal flows of information. This control via the information vector is becoming more granular, working now at the level of individual laborers rather than subcontracted firms. At first, the vectoral made capitalist firms subordinate. Now, where they can, the vectoralist class replaces them altogether with individual subcontractors.

Like all previous extensions of the abstraction of private property, this one too produces its own internal antagonist. And like all previous antagonists, it never appears in a pure and self-conscious form. Most peasants tugged the forelock and did what they were told, silently cursing the lord under their breath. Most workers settled for some job security and a weekend. Radical class-based movements are rare.

So it comes as no surprise that the hacker class is not particularly conscious or organized or antagonistic either. But its frustrations are real. The hacker class designs the information tools by which all human effort is controlled and organized by asymmetrical flows of information. The hacker can see her or his own job succumbing to this tendency in the end as well.

The organization of the activity of hackers is built into the form of code itself. Their efforts are compartmentalized and separated—blackboxed. They work on alienated tasks just as workers do. Only they do not work from clock-on to knock-off time. Even when they sleep they work for the boss. They might in some cases be well paid, but in many instances they are not. Their skills date quickly, and they are replaced by others.

Hackers won't necessarily respond to the vectoralist class in traditional labor movement terms. A strike would hardly be effective given that hackers can't shut down production. The most frequent forms of antagonism are more likely changing jobs, or stealing time on the job for one's own projects. Of course many dream of start-up glory, but that dream quickly tarnishes when the hacker gets to see firsthand who usually cashes out first in such schemes.

The significance of platform cooperativism is that it is a movement that can place itself at the nexus of the interests and experiences of both workers and hackers. Why not use the specific skills hackers have to create the means of organizing information, but use it to create quite other ways of organizing labor? Cooperatives have a long history in the labor movement; indeed, in their origins, they looked back to forms of peasant self-organization of the commons.

Why not re-imagine the cooperative on the basis of contemporary forms of information vector—but without the information asymmetries that are the basis of vectoralist class power? That seems like the thread of a political-economic project that both honors past struggles and also addresses the distinctive form of commodification in the age of the information vector as a private property relation.

The vectoral political economy is in many ways worse than the capitalist one. It gives the ruling class of our time unprecedented wealth amid growing poverty and despoliation. It enables that ruling class unprecedented flexibility in routing around strikes, blockages, or communal strongholds. It has made the whole planet appear as an infinitely exploitable resource at precisely the moment when it is also clear that the past products of commodified production are coming back to haunt us.

And yet every advance in the abstraction of the form of private property also opens up new perspectives on what may be held in common, and how the common might counter-organize. The practical and conceptual experiments of platform cooperativism are a key moment in the advance of this counter-organizing agenda.

9. HOW THE UN-SHARING ECONOMY THREATENS WORKERS

STEVEN HILL

The U.S. workforce has been enduring a long downward spiral for nearly three decades. That's how long it's been since American workers, in aggregate, have had a pay increase. Even as corporate profits are at an all-time high, with significant chunks of it parked overseas in tax havens to avoid being taxed, not many of the benefits of that labor productivity are being returned to domestic shores.

A significant factor in the decline of the quality of jobs today has been the increasing reliance by many employers on "non-regular" employees—a growing army of contractors, freelancers, temps, and part-timers that form the precarious vanguard in a "freelance society." Any proposal to revamp platform capitalism and launch platform cooperativism is challenged with coming to grips with how technology is changing the nature of work across an astonishing range of occupations and industries.

Meet Chris Young, an assembly-line worker at a Nissan manufacturing plant in Tennessee. Young works alongside other Nissan employees, but he works for a private contractor who now provides a majority of Nissan's workers. Young receives half the salary, less job security, and fewer safety-net benefits than the regular Nissan employees, even though he does the exact same job.

Auto manufacturers increasingly rely on this kind of two-tiered system. Nationwide, temps have provided nearly a fifth of the job growth since the recession ended. And increasingly, the temps aren't very temporary. Some have been employed at the same company for as long as eleven years, resulting in the doublespeak term "perma-temps." Microsoft paid $97 million to settle a lawsuit for denying benefits to over eight thousand perma-temps.

Besides temp workers, another type of worker is known as the "independent contractor." Fritz Elienberg worked for five years as a full-time employee installing cable and Internet service for RCN Corporation in Boston. Elienberg often worked ten to fourteen hours a day yet he never received time-and-a-half for overtime. When a ladder fell on his foot and seriously injured it, workers' compensation would not cover his medical bills. Why? Because RCN did not regard him as a regular employee; instead he was an "independent contractor." That meant, legally speaking, he worked for himself and was not employed by RCN. Elienberg sued RCN and the company promptly fired him, adding retaliation to his list of grievances.

The business advantage of using such non-regular workers is obvious: it can lower labor costs by 30 percent, since the business is not responsible for providing health benefits, Social Security, unemployment or injured workers' compensation, paid sick leave or vacation, and more. Contract workers, who are barred from forming labor unions and have no grievance procedure, can be dismissed without notice. A small percentage of contract workers, especially in the tech industry, earn high enough wages to make it all tenable, but most are helpless tumbleweeds in the erratic labor market of the freelance society.

Besides the explosion in the number of temporary and contract jobs, nearly half of the new jobs created in the so-called "recovery" pay only a bit more than minimum wage. Three-fourths of Americans are living paycheck to paycheck, with little to no emergency savings to rely on if they lose their job. Income inequality is now as bad as it was in 1928, just before the Great Depression. Incredibly, the share of wealth held by the bottom 90 percent is no higher today than during our grandparents' time. It's as if the New Deal had never existed.

RACE TO THE BOTTOM IN THE FREELANCE SOCIETY

Now a new and alarming mash-up of Silicon Valley technology and Wall Street greed is thrusting upon us the latest economic trend: the so-called sharing (or gig) economy. Companies like Uber, Instacart, Upwork, and TaskRabbit allegedly are "liberating workers" to become "independent entrepreneurs" and the "CEOs of their own businesses." In reality, these workers also are contractors, with little choice but to hire themselves out for ever-smaller jobs ("gigs") at low wages and with no safety net, while the companies profit.

Silicon Valley is redesigning the corporation itself. These gig companies are little more than a website and an app, with a small number of executives and regular employees who oversee an army of freelancers, temps, and contractors. In the vision of the techno gurus and their Ayn Rand libertarianism, CEOs want a labor force they can turn off and on like the latest Netflix movie.

For example, Upwork is an online business portal that acts like an eBay for jobs, allowing each worker to hang out a shingle to attract buyers of their services. A mere eight hundred employees (two-thirds of whom are contractors) oversee an army of ten million freelancers from all over the world who compete against each other, scrounging for jobs in an online labor auction in which the bidders offering the lowest wages usually win. The types of jobs on the auction block include website and app designers, software developers, logo and graphic designers, translators, architects, engineers, and more. Workers from India, Thailand, and the developing world compete against developed world workers, undercutting each other's wages. It's a race to the bottom.

As contractors, these workers don't receive safety-net benefits because they aren't "employees" of whoever hires them. They also aren't paid while they are hustling for their next gig, a never-ending activity. Increasingly, these sorts of online job brokerages comprise a bigger chunk of the overall work force. TaskRabbit, CrowdFlower, Work Market, HourlyNerd (for hiring freelance MBAs), Thumbtack, and Freelancer.com (all of which either currently or in the past have used a similar online auction price structure) have infinitely expanded the geographic range and size of the contingent labor job applicant pool.

With wages flat, the quality of jobs declining, and the safety net deteriorating, a whiff of desperation has crept into the labor force. Businesses large and small, whether in the traditional economy or the sharing economy, are gradually distancing themselves from any enduring relationship with the workers they hire. "Fissured" work is an increasingly common feature of our outsourced economy. It's how more and more people are working, whether they want to or not. Welcome to the Freelance Society.

THE UBER-IZATION OF WORK

Uber is the best known of these new kinds of businesses. It is nothing more than a temp agency, in which the predominant job on offer is that of a taxi driver (more recently Uber is trying other related services, such as courier or delivery person). Drivers are not treated as employees but as freelance contractors, and most drivers, after they subtract their considerable driving expenses, don't earn any more than taxis drivers. Indeed, many Uber drivers complain they don't earn minimum wage, much less a living wage. They receive no safety-net benefits and can be cut off the app-based platform at any time. Recently Uber cut off hundreds of drivers (and possibly over a thousand) in Los Angeles and San Francisco because those drivers' "acceptance rate" was too low. Many veteran drivers have figured out that, given the increase in congestion (in part stemming from the proliferation of ride-sharing vehicles on the streets), drivers often don't make any money on short rides because they get stuck in traffic. They have begun refusing short rides, so Uber fired drivers it deemed to be offending without warning.

If these workers really are the "CEOs of their own driving business," as Uber likes to claim, shouldn't they be able to refuse a ride they know will cause them to lose money? This incident and others seem to support the legal claim by thousands of drivers who are suing Uber, insisting they are indeed employees under the strict management of Uber, not sovereign contractors. As an employer, Uber would be responsible for paying Social Security and Medicare contributions for these workers, as well as unemployment and injured workers' compensation and driving expenses. This stark reality also points to the grave need for the creation of a new ride-sharing platform in which drivers have more control, either through outright cooperative ownership of the platform or a binding contract negotiated by a union-type organization, such as was recently empowered in Seattle via new legislation.

According to Uber's own numbers, most drivers work only part-time and leave after a year. New drivers like the flexibility, but after a while they burn out, with frequent wage cuts and unfair treatment. In January 2016, Uber slashed wages once again, this time by 30 percent to about 50 cents per mile in some locations (after Uber's 25 percent cut of each fare is subtracted). If driving for Uber was such a great job and paid halfway decently, wouldn't more drivers last longer and drive more hours?

Many businesses are increasingly relying on these types of operations as a core part of their profit-maximizing model. If this new corporate model is left unregulated, it will destroy what remains of a vibrant middle class. But fortunately there are solutions. One that I and others have proposed is creating a "universal and portable safety net." Each worker should be assigned an "Individual Security Account" into which every business that hires that worker would pay a small "safety net fee," prorated to the number of hours a worker is employed by that business. Those funds would be used to pay for each worker's safety net.

We don't have to wait for a dysfunctional U.S. Congress to pass this new kind of deal. State governments and even city councils can pass it, requiring local businesses to pay into Individual Security Accounts for each worker. By modernizing the social contract, and combining that with the creation of cooperative platforms and greater economic democracy that can offer to consumers an alternative to runaway capitalism, we would take major steps toward forging a world in which most workers would be enriched by technology and innovation, instead of being disrupted and deprived by a freelance society and its "share-the-crumbs" economy.

10. SPONGEBOB, WHY DON'T YOU WORK HARDER?

CHRISTOPH SPEHR

As we head once again into a brave new world, this time character-ized by buzzwords such as "platform" and "mutual" and "decen-tered," it is worthwhile to remember the fundamental difference between the cordless drill and emancipation. Cordless drills came into the world as a spin-off—of space travel, from there spreading through earthly households and construction sites. Emancipation, especially the emancipation of labor, does not. It does not emerge as a by-product of technological development, and this still holds true in the age of platforms and algorithmic capitalism. Yes, there are great possibilities for redefining the position of labor in the production process. Yes, the force is strong in the new means of production, but so is the dark side.

Should you feel discouraged from opening up a new plat-form cooperative? No, not at all. But do not expect exploitative, hierarchical, narrow-minded capitalism to roll over and die just because you're clever enough to program a platform of your own. A new mode of production that could release the potential for a less alienated, less exploited, less asshole-infested worklife will not prevail just because it seems economically superior. Merely embracing the change will not be good enough. You can't win by just being better. You have to change the rules, which implies getting organized.

WORKING HARDER

SpongeBob: Mr. Krabs, you wanted to ask me a question.
Mr. Krabs: Yes, SpongeBob. Why don't you work harder?
SpongeBob: I don't know, Mr. Krabs. I don't know!

To keep it brief and simple, listen to the master of brief and simple: SpongeBob SquarePants. In the landmark episode "Imitation Krabs," the shortest possible explanation of the capitalist-worker relation is offered when Mr. Krabs poses his most pressing concern: "Spongebob, why don't you work harder?"

This is what capitalism is all about: the capitalist (Mr. Krabs) buys and pays for the worker's (SpongeBob's) manpower but owns all of the product. The harder the worker works, the greater the difference between wage and creation of surplus value, the greater the profit. The bulk of surplus goes to the pockets of the capitalist, enhancing his lead in social power. Simple, right?

A common method for the capitalist to increase value creation is handing the worker a tool (in SpongeBob's case, the spatula and the secret recipe), enabling the worker to be more productive. The worker is dependent on the means of production, which are owned by the capitalist, because the worker can only be productive through them. Yes, you own a laptop and a mobile phone now; but in analyzing a mode of production, the decisive question is not who owns *any* kind of means of production but who owns the *dominant* means of production. These used to be the factory, the machinery. They are now becoming the big algorithms, the constantly adjusted and ever-developing virtual machinery. If you own them, you're the capitalist. If someone else owns the majority of your company because you needed investors, or if you depend on the platforms and algorithms owned by others, you're not the capitalist. (You'd "get cash, but they'd get the reins," as Nathan Schneider puts it.)

PAYING TO BREATHE

Mr. Krabs: Breathe on your own time. I don't pay you to breathe!
Squidward: (unfolding his pay slip) What is this? You want me to pay for
standing at the cash register?
Mr. Krabs: There's gonna be a few changes around here.

As passionate SpongeBob fans know, things improve very little in the Krusty Krab. Even the fabulous turbo-drive spatula from the episode "Help Wanted" never reappears once the "anchovies situation" is over. Capital can be quite reluctant to push for progress. It costs a lot of money, and once everybody has it, the gains in productivity go to the consumer.

Throughout the history of capitalism, capitalists used to increase their profits through changes that do not improve the means of production but shift costs and burdens from the capitalist to the worker or the society—for example, by claiming something as capitalist property that used to be common property. In nineteenth-century England it was land. In the twenty-first-century global economy it is information appropriated by Facebook, Google, and Microsoft. In proto-industrialization, the cottage system (when the workers had to work at home) allowed the manufacturer maximum flexibility, minimum responsibility, and low wages. Today, ubiquitous platformization does the same. In many cases, its competitiveness stems only from de-organizing and de-valuing labor.

This is not progress. According to Stephen Marglin (in his famous essay "What Do Bosses Do?"), a method of production is technologically superior ("progressive") only if it produces the same output with less input. There is no economic superiority in producing something cheaper just by paying lower wages, by skimming labor costs through unpaid hours or unrewarded density of work, or by making the worker pay for breathing.

Nevertheless, there is such a thing as progress. It happens even in capitalism, mostly when labor is organized and society is on alert for externalization of costs (so that other exits are closed). While competition may work as an incentive to apply progress, it is produced outside competition: through cooperation, in spaces of non-efficiency.

CHANGING THE RULES

> *SpongeBob: We are workers united! We're gonna smash that with the people's hammer! And we're gonna...wait...Squidward, what was that other thing?*
> *Squidward: Dismantle the oppressive system.*
> *SpongeBob: Yeah, that one, too!*

Ah, what was that other thing I mentioned above? Wait...yes: you have to close the exits!

To push for economic transformation, for the cooperatives to take over, you have to define the game as ruled by progress. And as we have learned, this means closing the exits, preventing the possibility of thriving not by progress but by shifting burdens and costs to workers and society. This cannot be done individually. Rules are collective, and obtaining a change of rules means getting organized.

What we have to obtain is a bill of rights for the age of algorithmic capitalism. First, we need a fight for new labor rights to force back the new forms of exploitation that are running wild. Workers have a right to know their fellow workers, especially in platform labor. They have a right to balanced job packages; to a fixed space of self-determined labor as part of their job (expanding Google's "20 percent time" to all workers); to rules-based, democratic governance instead of command-based management. They have a right to a share in productivity gains; a right to log off; a right to work without surveillance. That's not only trade-union work. Palak Shah's "Good Work Code" or Trebor Scholz' principles for platform co-ops are today's equivalent to the demand for an eight-hour day.

Second, we need new entrepreneurial rights—today's equivalent to the antitrust laws—to fight back the stifling power of size, financial markets, and incumbency. Network neutrality and equal rights for cooperatives are essential, as is equal access to supply, services, publicity, and clouds. Enterprises need a right to set up their own constitutions and to represent their interests in the context of holdings and investors. As Nathan Schneider writes, "A new economy will need new public politics to level the playing field between traditional corporations and collaborative enterprises."

Third, we need a legal framework for a new regime of accumulation. This takes vision; it's today's equivalent of what a social-democratic or socialist economic policy used to be. It means bringing state and society back in, not only as neutral gatekeepers of economic fair play but as a *volonté générale* that gives the economy a social purpose and a base in democratic values. There is nothing neutral about the actual economy. It is a complicated, systematized effort to reconcile productivity with the privilege of powerful elites, dominant social groups, and global coalitions. A democratic, cooperative economy will likewise be a systematized effort to reconcile productivity with equality, sustainability, and "the liberation of the productivity of all" (as Bertolt Brecht stated in his lines on the "great production" in his journals). This might mean public investment, public co-ownership, and strong incentives for social enterprises.

Some of this might sound slightly awkward to us, since we haven't discussed it for a long time. But we have to have this conversation if we want to implement cooperativism. I fully agree with Trebor Scholz when he states: "This isn't about some romantic attachment to the past. This is about the language of labor and living within it, its cardinal lesson, which is that *in confrontation with the power of the employing class, individual solutions are not working.*" Or, as one person in the audience at the Platform Cooperativism conference put it: "Please shut up and grow some class consciousness."

11. PORTABLE REPUTATION IN THE ON-DEMAND ECONOMY

KATI SIPP

While the app-based gig economy derives a certain sexiness from its association with the tech world, the gig economy has existed offline for generations. Some workers have been pushed into the gig economy by circumstances beyond their control, while others have always chosen and will continue to choose it, either due to the nature of their occupations or for personal reasons. Workers in gig situations get new jobs through the strength of their reputations; the difference in the on-demand economy is that workers don't own their reputations.

"Just like domestic workers were tucked away in people's houses, digital laborers remain invisible, tucked away in between algorithms," said Trebor Scholz in his opening talk at the Platform Cooperativism conference. This observation resonated with me, because I worked for many years with homecare workers and their allies in the disability rights movement, who have the slogan "Invisible No More." One of the big differences between domestic workers and platform workers (especially online-only platforms like Amazon's Mechanical Turk) is that the domestic workers have word of mouth, and references, and do work that is geographically located in a specific place.

Most of us employed in traditional jobs have things like resumes, coworkers that we can rely on for references, networks of people who will tell us about jobs, and other advantages that help when we're looking for new work. I personally have changed jobs twice in the past two years, and both times I took my reputation with me—through word of mouth, through work-related networks, and yes, through

LinkedIn recommendations. In some ways, the platform economy has the potential to be wildly democratizing, because more transparent networks for finding work should mean larger numbers of people getting new opportunities.

Many of these platforms don't let workers have any control over their reputations. I don't want to sugarcoat the problems of reputation for workers with traditional jobs, but in some ways reputation is much more punishing for platform workers. There have been many stories about Airbnb, Uber, and others removing workers from their platforms, with little to no notice or ability to correct problems. In fact, Uber drivers are required to maintain a certain rating in order to stay on the platform—a fact that few passengers know. Workers in most cases lack the ability to challenge the stain on their reputations, and sometimes they don't even know why their reputations might have suffered. Platforms are highly dependent on customer ratings for policing the quality of their workforce, but they haven't figured out how to correct for those same customers' race and gender biases. It can feel to the worker like it's "one strike and you're out"—and that arbitrariness just adds to the instability of gig work. In addition, reputation isn't portable. If Uber drivers want to change platforms and start delivering packages for Instacart, they have to start from scratch to build up a good reputation on the new site—even though they are using skills that are valuable to both sites.

It doesn't have to be this way, and the offline gig economy reminds us of that.

Meet my friend Dave. Dave is an actor who lives in New York City. The nature of acting is contingent—even the longest-running Broadway musical will come to an end long before the end of an individual actor's career. The same is true for movies, TV shows, commercial work, recording audio books, and Shakespeare in the Park (or, in his case, Shakespeare in the Parking Lot).

Like most gig workers, Dave has multiple sources of income throughout the year—and to supplement them, he also works as a catering bartender. In years when he books more acting work, he might bartend less, but he's been a steady bartender for a long time. Catering is itself seasonal work. There's lots to do in December, when rich people or companies are having their holiday parties, but much less in January. Catering managers know him, and they know he's reliable. They keep calling him for jobs, even in years when he might be turning them down frequently for acting gigs.

For Dave, one part of the solution to his "gig" insecurity is his union, the Screen Actors Guild–American Federation of Television and Radio Artists, or SAG-AFTRA. The union negotiates multi-employer agreements within different industry sectors—so the contract for TV commercial work in New York is different from the contract for recording audio books—but all the employers pay into common health, benefits, and pension funds, on a per capita basis. If Dave books a one-day commercial shoot, the ad agency pays less into the pension than if he books a recurring role on a TV show, of course, but all the money goes into one pension that Dave will eventually be able to retire on.

It's important for folks in the platform-cooperative community to understand that there are existing worker-led organizations that are set up to deal with multi-employer, disaggregated work situations—and that we can build from their model, rather than starting from scratch. Not every union was set up to deal with jobs in which workers stayed with the same company for years on end. Lots of people are starting to think about ways that workers can organize in the gig economy—and I want to urge all of them to think about how to build reputation best-practices into their efforts. A decent reputation system should be:

- Worker-controlled
- Transparent
- Reparable
- Able to take input from multiple companies (if that is relevant to the worker)
- Resistant to bias and prejudice
- Fair in how it distributes rewards

- Correctable with improved behavior
- Equipped with some kind of a grievance process

Dave's union was formed the old-fashioned way. Actors struck in Hollywood in 1929, and then radio producers in 1939. The technological tools that actors have worked with have changed over the years—and the union has transitioned with the new technology. Now, for instance, it is organizing workers who voice video games, figuring out how to deal with new channels for distribution of content like YouTube and Hulu, and taking on the digital advertising industry. SAG-AFTRA hasn't had to "solve" the reputation problem in the way that online platform worker organizations will—but that is because they are dealing with employers who want to see actors' work through auditions before they cast them. We don't get to ask to see samples of our TaskRabbit's work assembling Ikea furniture; we choose them based on their on-site ratings instead. Dave still relies on reputation in his catering work, but that's delivered through his existing (and offline) personal network.

We are already seeing tech companies develop ways of aggregating our online reputations—through sites like LinkedIn, Karma, MakerBase, and Work Hands—but those platforms still haven't caught up to the best-practices of the offline gig economy. And let's be honest; you wouldn't want a union that just exists to protect workers' reputations, just as we don't organize offline unions only around issues of worker reputation.

Those of us who are striving to organize workers in the online economy have to build a theory for reputation portability and protection into our other organizing work. We can't let reputation management become disaggregated from the platforms on which workers get work. So build a better mousetrap. We should take a lesson from Dave's union too, and build organizations that can evolve as the technology work evolves.

12. COUNTERANTI-DISINTERMEDIATION

DMYTRI KLEINER

In Chapter 33 of *Capital*, Karl Marx introduces us to the character of Mr. Peel, recounted from E. G. Wakefield's *England and America: A Comparison of the Social and Political State of Both Nations*. Although Mr. Peel's story is one of early nineteenth-century colonialism, it helps us understand what has become of the Internet and the so-called sharing economy.

Mr. Peel went to Swan River in Australia to seek his fortune. He brought everything an aspiring capitalist might need to start accumulating surplus value and become a great capitalist: three hundred people, including men, women, and children, to provide the labor and its reproduction, along with £50,000, a large sum at the time.

However, things didn't work out for Mr. Peel, as Marx concludes, "Unhappy Mr. Peel who provided for everything except the export of English modes of production to Swan River!"

Once arrived in Swan River, the three hundred people simply went off and settled on the vast amounts of free land available, and "Mr. Peel was left without a servant to make his bed or fetch him water from the river."

He discovered that capital is not a thing but a social relation between persons, established by the instrumentality of things.

As Marx explains further, "Property in money, means of subsistence, machines, and other means of production, does not as yet stamp a man as a capitalist if there be wanting the correlative—the wage-worker, the other man who is compelled to sell himself of his own free will."

Marx argues, "The means of production and subsistence, while they remain the property of the immediate producer, are not capital. They become capital only under circumstances in which they serve at the same time as means of exploitation and subjection of the laborer."

Mr. Peel's capitalist class was not satisfied with their inability to expand their mode of production into the colonies, and found a solution in enclosure, described by Wakefield as "Systematic Colonialization."

Land was seized by law as public property and privatized, with no free land available. Only those with wealth could be owners, and thus everybody else needed to sell their labor to capitalists.

The early Internet was like Swan River. How can the modern Mr. Peel make money operating Internet platforms, if anybody can do so? If all the software and the networks were open and widely available, then nobody could really make significant profit. If the means of production are available to all, then there can be no capital.

Like the colonies, the Internet needed to be systematically colonized in order to create the conditions needed by capital. This was also accomplished by enclosure. The original infrastructure was taken over and brought under capital control, and decentralized systems were displaced with centralized systems.

"Social media" and "sharing" platforms are two forms of this centralization, two business models for platform capitalism.

SURPLUS VALUE VS. SURPLUS PROFIT

It's tempting to look at sites like Facebook and YouTube and conclude that they earn their profits by exploiting their own users, who generate all the content that makes the sites popular. However, this is not the case, because the media is not sold and therefore makes no profit and captures no value.

What is sold is advertisement, thus the paying customers are the advertisers, and what is being sold are the users themselves, not their content.

This means that the source of value that becomes Facebook's profits is the work done by the workers in the global fields and factories, who are producing the commodities being advertised to Facebook's audience.

The profits of the media monopolies are formed after surplus value has already been extracted. Their users are not exploited, but subjected, captured as audience, and instrumentalized to extract surplus profits from other sectors of the ownership class.

Sharing economy companies like Uber and Airbnb, which own no vehicles or real-estate, capture profits from the operators of the cars and apartments for which they provide the marketplace.

Neither of these business models is very new. Media businesses selling audience commodity are at least as old as commercial radio. Marketplace landlords, capturing rents from market vendors, have been with us for centuries.

Rather than subvert capitalism, "sharing" platforms have been captured by it.

CONSENT-ORIENTED ARCHITECTURE

Capitalist platforms based on the sale of audience commodity and capturing marketplace rents demand a sacrifice of privacy and autonomy.

Audience commodity, like all commodities, is sold by measure and grade. Eggs are sold in dozens as grade "A." An advertiser might buy audience commodity by thousands of clicks from middle-aged white men who own a car and have a good credit rating with a certain measure—e.g., 10,000 clicks.

Audience commodity is graded by what is known about the audience's demographics. Platforms with business models that sell audience commodity require surveillance. Likewise, platforms that capture marketplace rents collect extensive data on their users and providers in order to maximize the profitability of the platform.

A mandatory sacrifice of consent is required to use the platforms. When a user shares information on the platform, they may consent to sharing that information with certain people, but they don't necessarily consent to that information being available to the platform's staff, to advertisers, or to business partners and state intelligence. Yet for most users there are no practical alternatives, and they must sacrifice such consent in order to use the platform.

Corporations built to maximize profits are unable to build consensual platforms. Their business model depends fundamentally on surveillance and behavioral control.

True consensual platforms should have privacy, security, and anonymity as core features.

The most effective way to ensure consent is to ensure that all user data and control of all user interaction resides with the software running on the user's own computer, and not on any intermediary servers.

COUNTERANTIDISINTERMEDIATION

On her blog, Wendy M. Grossman writes, "'Disintermediation' was one of the buzzwords of the early 1990s. The Net was going to eliminate middlemen by allowing us all to deal with each other directly." Today, the landscape is dominated by many fewer, much larger ISPs whose fixed connections are far more trackable and controllable. We thought a lot about encryption as a protector of privacy and, I now think, not enough about the unprecedented potential for endemic wiretapping that would be enabled by an increasingly centralized Internet.

The idea of disintermediation was central to the emancipatory visions of the Internet, yet the landscape today is more mediated than ever before. If we want to think more about the consequences of an increasingly centralized Internet, we need to start by addressing the cause of this centralizing. The Internet was colonized by capitalist platforms; centralization is required to capture profit. Disintermediating platforms were ultimately reintermediated by capitalist investors dictating that communications systems be built to capture profit.

The flaw was, to some degree, a result of the architecture of the early Internet. The systems that people used in the early Internet were mainly cooperative and decentralized, but they were not end-to-end. Users of email services and Usenet, the two most used platforms, did not generally operate their own servers, on their own local computers, but were dependent on servers run by others.

Servers require upkeep. Operators need to finance hosting and administration. As the Internet grew beyond its relatively small early base, Internet service came to be provided by capitalist corporations, rather than public institutions, small businesses, or universities. Open, decentralized services came to be replaced by private, centralized platforms. The profit interests of the platform financiers resulted in a policy of antidisintermediation.

Just as systematic colonialization was developed to establish the capitalist mode of production in the colonies, antidisintermediation was developed to colonize cyberspace.

The basic strategy of antidisintermediation was formulated by economists like Carl Shapiro and Hal R. Varian. Their influential book *Information Rules* encourages platform owners to pursue "lock-in." As they summarize on their website, "Since information technology products work in systems, switching any single product can cost users dearly. The lock-in that results from such switching costs confers a huge competitive advantage to firms that manage their installed base of customers effectively."

Their advice was well received. Varian is currently chief economist of Google, while Shapiro was a deputy assistant attorney general for economics in the Department of Justice.

Going back to an early Internet architecture of cooperative, decentralized servers, as projects like Diaspora, GNU Social, and others are attempting to do, will not work. This is precisely the sort of architecture that antidisintermediation was designed to defeat.

Decentralized systems need to be designed to be counterantidisintermediationist.

Central to the counterantidisintermediationist design is the end-to-end principle; platforms must not depend on servers and admins, even when cooperatively run, but must, to the greatest degree possible, run on the computers of the platforms users.

The computational capacity and network access of the users' own computers must make up the resources of the platform, so that on average each new users adds more resources to the platform than they consume.

By keeping the computational capacity in the hands of the users, we prevent the communication platform from becoming capital, and we prevent the users from being instrumentalized as audience commodity.

Thus, we leave Mr. Peel just as unhappy in cyberspace as he was in Swan River, resisting the colonization of our communication platforms by venture capital and paving the way for venture communism.

13. FROM OPEN ACCESS TO DIGITAL COMMONS

DAVID BOLLIER

We are accustomed to regarding open platforms as synonymous with greater freedom and innovation. But as we have seen with the rise of Google, Facebook, and other tech giants, open platforms that are dominated by large corporations are only "free" within the boundaries of market norms and extractive business models. Yes, open platforms provide many valuable services at no (monetary) cost to users. But when some good or service is offered at no cost, it really means that the user is the product. In this case, our personal data, attention, social attitudes, lifestyle behavior, and even our digital identities are the commodity to which platform owners are seeking unrestricted access.

In this sense, many open platforms are not so benign. Many of them are techno-economic fortresses, bolstered by structural dynamics that enable dominant corporate players to monopolize and monetize a given sector of online activity. Market power based on such platforms can then be used to carry out surveillance of users' lives; erect barriers to open interoperability and sharing, sometimes in anti-competitive ways; and quietly manipulate the content and experience that users may have on such platforms.

Such outcomes on seemingly open platforms should not be entirely surprising; they represent the familiar quest of capitalist markets to engineer the acquisition of exclusive assets and mine them for private gain. The quarry in this case is our consciousness, creativity, and culture. The more forward-looking segments of capital realize that owning a platform (with stipulated, but undecipherable, terms of use) can be far more lucrative than owning exclusive intellectual property rights for content.

So for those of us who care about freedom in an elemental human and civic sense—beyond the narrow mercantilist "freedoms" offered by capitalist markets—the critical question is how to preserve certain inalienable human freedoms and shared cultural spaces. Can our free speech, freedom of association, and freedom to innovate flourish if the dominant network venues must first satisfy the demands of investors, corporate boards, and market metrics?

If we are serious about protecting human freedoms that have a life beyond markets, I believe we must begin to develop new modes of platform cooperativism that go beyond standard forms of corporate control. We need to pioneer technical, organizational, and financial forms that enable users to mutualize the benefits of their own online sharing. We must be able to avoid the coerced and undisclosed surrender of personal information and digital identity to third-parties who may or may not be reliable stewards of such information.

There are other reasons to move to commons-based platforms. As David P. Reed showed in a seminal 1999 paper, "That Sneaky Exponential," the value generated by networks increases exponentially as interactions move from a broadcasting model based on "best content" to a network of peer-to-peer transactions. The most valuable networks, however, are those that facilitate group affiliations to pursue shared goals—which is to say, networks that are treated like commons.

Reed found that the value of such "group forming networks," in which people have the tools for "free and responsible association for common purposes," to be 2^n, where n is the number of members in the network. That's a fantastically powerful growth curve. His analysis suggests that the value generated by Facebook, Twitter, and other proprietary network platforms remains highly rudimentary because participants have only limited tools for developing trust and confidence in each other. In short, the value potential of the commons has been deliberately stifled as part of the business model.

For all of these reasons, our imaginations and aspirations must begin to shift their focus from open platforms to digital commons. Self-organized commoners must be able to control the terms of their interactions and governance, and to reap the fruits of their own collaboration and sharing.

From open access platforms to managed digital commons: that is one of the chief challenges that network-based peer production must meet if we are going to unleash the enormous value that distributed, autonomous production can create.

A variety of legal and technological innovations are now starting to address the structural limits of (market-financed) open platforms as vehicles for commoning. These initiatives remain somewhat emergent, yet they are filled with great promise.

THE POTENTIAL OF THE BLOCKCHAIN

One instrument for converting open platforms into digital commons is the blockchain ledger, the software innovation that lies at the heart of the Bitcoin digital currency network. Although Bitcoin itself has been designed to serve familiar capitalist functions (tax avoidance, private accumulation through speculation), the blockchain ledger is significant because it can enable highly reliable, versatile forms of collective action on open networks. It does this by validating the authenticity of a digital object (for example, a bitcoin) without the need for a third-party guarantor such as a bank or government body.

This solves a particularly difficult collective-action problem in an open network context: How do you know that a given digital object—a bitcoin, a legal document, digital certificate, dataset, a vote, or a digital identity asserted by an individual—is the real thing and not a forgery? By using a searchable online ledger that keeps track of all transactions, blockchain technology solves this problem by acting as a kind of permanent record maintained by a vast, distributed peer network. This makes it far more secure than data kept at a centralized location, because the authenticity of its records are registered among so many nodes in the network that it is virtually impossible to corrupt.

Because of these capabilities, blockchain technology could provide a critical infrastructure for building what are called "distributed collaborative organizations" (sometimes called "distributed autonomous organizations"). These are essentially self-organized online commons. A DCO could use blockchain technology to give its members specified rights within the organization, which could be managed and guaranteed by the blockchain. These rights, in turn, could be linked to the conventional legal system to make the rights legally cognizable and enforceable.

One rudimentary example of how the blockchain might be used to facilitate a commons: In the United States, former Federal Communications Commission Chairman Reed Hundt has proposed using blockchain technology to create distributed networks of solar power on residential houses coordinated as commons. The ledger would keep track of how much energy a given homeowner generates and shares with others, and consumes. In effect, the system would enable the efficient organization of decentralized solar grids, together with a "green currency" that could serve as a medium of exchange within solar microgrids or networks, helping to propel adoption of solar panels. The blockchain amounts to a network-based architecture for enabling commons-based governance.

SMART TRANSACTIONS

This field of experimentation may yield another breakthrough tool for forging digital commons: smart contracts. These are dynamic software modules operating in an architecture of shared protocols (much like TCP/IP or HTTP) that could enable new types of group governance, decision-making, and rules-enforcement on open network platforms.

We are already familiar with rudimentary—and corporate-oriented versions—of this idea, such as digital rights management, a system that gives companies the ability to constrain how users may use their legally purchased technologies, from movies on DVD to ebooks. As the power of networked collaboration has become clear, however, many tech innovators now recognize that the real challenge is not how to lock up and privatize digital artifacts, but how to assure that they can be reliably shared on open platforms in legally enforceable ways, for the benefit of a defined group of contributors or for everyone.

There are now many active efforts underway to devise technical systems for deploying "smart" legal agents whose transactions would also be enforceable under conventional law. These transactions could, of course, be used to invent new types of markets, but they also could be used to create new types of commons. Ultimately, the two realms may bleed into each other and create social hybrids that conjoin community commitments and market activity.

A related realm of software innovation is trying to blend familiar cooperative structures with open network platforms to enable collective deliberation and governance—"commoning"—through online systems. Some of the more notable experiments include Loomio, DemocracyOS, and LiquidFeedback. Each of these seeks to enable members of online networks to carry on direct, sustained, and somewhat complicated discussions, and then to clarify group sentiment and reach decisions that participants see as binding, legitimate, and meaningful.

NETWORKS OF PEER PRODUCERS

In a natural extension of such capacities, open value networks, or OVNs, are attempts to enable bounded networks of participants to carry out crowdfunding, crowdsourcing of knowledge, and co-budgeting among their identifiable participants. OVNs such as Enspiral and Sensorica have been described as an "operating system for a new kind of organization" and a "pilot project for the new economy." These enterprises consist of digital platforms that facilitate new modes of decentralized and self-organized social governance, production, and livelihoods among members of distinct communities. The networks are organized in ways that let anyone contribute to the project and be rewarded based on their contributions—as measured by actual contributions, experience, and other collectively determined criteria.

Unlike most commons, which are intended to serve household or community needs, not market gain (e.g., commons for water, urban spaces, open access publishing, FabLabs, and makerspaces), open value networks have no reservations about engaging with markets; users of OVNs simply wish to maintain their organizational and cultural integrity as commons-based peer producers. This means open, horizontal, and large-scale cooperation and coordination; responsible stewardship of the shared wealth and assets while allowing individual access, use, authorship, and ownership of resources where appropriate; careful accounting of individual inputs and outcomes via a common ledger system; and the distribution of fair rewards based on individual contributions to the project.

These initiatives to create new technical, organizational, and financing opportunities for platform cooperativism are still emerging. They will require further experimentation and development to make them fully functional and scalable. Yet they promise to furnish attractive, potentially breakthrough alternatives to centralized, profit-centric platforms. By providing more trustworthy systems for genuine commoning and user sovereignty, these new forms could soon enable digital commons—and hybrid forms of user-driven markets—to surpass the value-creating capacities of conventional open platforms.

PART 3

AN INTERNET OF OUR OWN

SHOWCASE 1: PLATFORMS

What would it be like to use a cooperative Internet? How would we interact with it differently? How would we protect our rights and meet our needs? These projects answer such questions in a variety of ways. While not all are formal co-ops, they replicate existing tools in fairer forms, in addition to imagining new possibilities that cooperation makes possible. Some are still little more than an idea, while others are earning millions of dollars in revenue. But they all demonstrate that platform cooperativism is under way already.

Project Name: Stocksy United
Completed by: Nuno Silva & Brianna Wettlaufer
Location: Victoria, British Columbia
URL: stocksy.com

Stocksy is a stock photo agency providing royalty-free licenses on exclusive photos via an online marketplace that provides sustainable careers to photographers through co-ownership, profit sharing, and transparent business practices. Our content is curated to challenge the status quo of stock photos. We're very selective of our photos and our members in order to provide a premium product at an accessible price-point.

Each member owns an equal voting share in the company. We're organized into three classes: founders and advisors, staff, and photographers. Daily operations are managed by executive staff and employees in a flat decision-making structure to encourage ownership and enthusiasm for each individual's contributions. Our board includes directors from each class. Any member can propose resolutions. Annual general meetings are held to report, discuss, and vote on the business and strategy.

We're growing at a sustainable and controllable rate. Our founders knew that growing too fast can often lead to distractions from core values. Marketing has been modest; we let our aesthetic and word of mouth be the most significant driving factor for acquisition. Tech and product development grow out of necessity with an understanding of what's required to remain competitive and with careful thinking about how things could be done better and more efficiently given our limited resources. Our greatest innovations have come from listening to our customers and knowing how to do more with less.

Growth has exceeded projections year after year. We have a great reputation in the industry as having a premium brand and enviable membership. Our members are seeing some of the highest revenues they've ever seen; they will receive annual dividends that account for 90 percent of the company's profit. In a few years we'll launch video licensing and begin developing an innovative search solution for finding the perfect image.

Project Name: Fairmondo
Completed by: Felix Weth
Location: Berlin, Germany
URL: fairmondo.com

Fairmondo is an online marketplace owned by its users. It is open to businesses as well as individuals, with no general restrictions on what products and services can be offered, except for illegal offers or offers deemed unacceptable by our members. By contrast, through the positive promotion of products that fulfill a set of criteria for "fairness," Fairmondo makes it easy for users to shop in line with their values. These criteria are constantly open for discussion and improvement by members and the broader user base.

Founded in Germany in 2012, Fairmondo is a multi-stakeholder cooperative with open membership for every person who feels affected by its activities. Its statutes include a legally binding commitment to uncompromising transparency and democratic accountability. The managing board is elected by the employees, to ensure a culture of mutual respect within the operating team.

In September 2013, we launched the German marketplace running on our self-developed open source platform. To start off, we focused on building a network of sellers to make the marketplace valuable to customers. Currently, it offers over two million products, the majority being books and media articles. For financing, we aimed at keeping the business 100 percent in the hands of the crowd. Over 2,000 members invested over €600,000 to make it happen. While this approach lends credibility, it also brings obvious challenges when it comes to scaling.

Our next step is pushing forward: bringing Fairmondo to other countries, in the form of autonomous co-ops owned by local users. Through sharing resources on technical development and outreach, we collaborate toward the goal of creating a true multinational cooperative, strong enough to challenge the big players in e-commerce.

Project Name: Coopify
Completed by: Steven Lee
Location: New York City
URL: coopify.org

The goal of Coopify is to empower and assist low-income worker cooperative members and entrepreneurs through the power of technology. It provides three key benefits. First, Coopify creates a brand for the marketplace designed to engage consumers to hire low-income members of cooperative businesses, such as home-care workers and movers. This, in turn, drives the second benefit: more work for existing cooperative members and the ability to onboard new members into the cooperative. More worker members means a faster path to sustainability and increased purchasing power for cooperatives. And, third, Coopify will simplify tasks for cooperative members. Currently, cooperative members have little to no technology tailored to their needs and often rely on manual organization and phone calls to book appointments, receive payment, attend trainings, and other tasks. Coopify eliminates these cumbersome processes by allowing members to interface with the platform in their native language (such as Spanish or Mandarin), respond to customer requests by texts, manage their own schedules, and receive payment—all on their smartphone.

Currently, the plan is to turn Coopify itself into a cooperative, with shared ownership and governance.

There are three priority areas. We need to build the platform. We need to sell the platform to potential cooperatives who might be willing to join. And we need to market the platform to potential consumers.

We're starting small with a community—cooperatives—that already has a built-in consumer base. We make their jobs easier and hope to bring in more consumers. Going forward, we'd like Coopify to be the go-to platform for New Yorkers to book services.

Project Name: Gratipay
Completed by: Chad Whitacre
Location: Global, with headquarters in Ambridge, Pennsylvania
URL: gratipay.com

Gratipay offers payments and payrolls for open organizations via the Internet. Our value proposition is that we enable open organizations to comply with the global financial and legal system. We compete on mission because we ourselves are an open organization, and we compete on cost because our fees are pay-what-you-want through our own platform.

As a legal entity, Gratipay is an LLC with a minimal set of owners, just enough to get by. As an open organization in the free and open-source software tradition, Gratipay is a benevolent dictatorship that shares power broadly through public, open decision-making on the Internet. Anyone willing to behave well is free to voluntarily collaborate in our work and share in our revenue. (See inside.gratipay.com for details.)

Gratipay is funded from revenue. Our interest in growing faster than revenue allows is offset by our difficulty fitting into traditional capitalist or philanthropic boxes, by wariness of greed, and by the example of successfully bootstrapped companies.

Gratipay has processed over $1 million since we launched four years ago. We're proud of this modest accomplishment, because we've achieved it while pioneering an open organization in a heavily regulated industry. We're even prouder of the way our open organization has enabled people to find not just economic support, but meaning and purpose in a voluntary community of work—a community that has already survived several existential threats together. Today, we process about $5,000 per month for about 150 projects and organizations. We don't know how big we're supposed to get, nor how fast we're supposed to get there. Our goal over the next few years is to stay faithful to our mission: to cultivate an economy of gratitude, generosity, and love.

Project Name: FairCoop
Completed by: Enric Duran
Location: Earth
URL: fair.coop

FairCoop is not about offering a specific service but building a full economic ecosystem for a postcapitalist society. In this sense, the ecosystem is what is going to be offered, offering individuals, collectives, cooperatives, and social companies a set of tools for connecting with and supporting each other and anyone aiming to make real, radical social change.

FairCoop does not have a legal entity at the moment, so there is not a legal basis for ownership. FairCoin is a peer-to-peer cryptocurrency based in free software, so in that sense nobody owns it; everyone who has some faircoins and runs the wallet software is part of the decentralized ownership of the currency system. Governance takes place in an open, participatory process through online assemblies every month and open discussions that can be accessed at the Fair.Coop site.

Openness is the main characteristic of the FairCoop development. Developers and activists with strong will are the main elements for building our initiatives and this community involvement and backing significantly helps financial costs. The cooperation and horizontal context are also important to take into account. The financial difficulties for the cooperative activity have been resisted partially using the FairCoin monetary hack, and we are organizing to add different mechanisms of economic disobedience to generate constant incomes for the common good.

FairCoop is still in early development stages, but step by step we are building the ecosystem. For example, FairMarket is an online marketplace. Aside from our main focus we hope to extend the local nodes network, which is the main infrastructure needed for deploying resources at the community level, specially the FairFunds, which consists of 20 percent of all the faircoins in circulation, and which will be distributed to commons-producing projects when the network becomes strong enough.

Project Name: Member's Media, Ltd. Cooperative
Completed by: Robert Benjamin
Location: Portland, Oregon
URL: membersmedia.net

Member's Media is the equivalent of the Studio System for independent film and TV. Our directive is to increase the quality, quantity, and value of independent narrative media, and give the audience a true voice in the creation of content that is produced for their consumption. Member's Media offers development and production, as well as support, to aspiring micro-budget filmmakers from diverse communities. The ultimate goal is to connect a large supporting audience with a slate of high-quality, independently produced narrative content through an online and mobile platform that is majority-owned by the users.

Member's Media is a multi-stakeholder Limited Cooperative Association or "Balanced Ownership Cooperative." There are four classes of patron owners (Supporter, Collaborator, Creator, and Mentor) and two classes of investor owners (Investor and Builder). The organization strives for the "golden mean," balance between the interests of each member class and the overall health of the cooperative. A comprehensive set of founding documents provide the framework for how the community interacts and supports each other's efforts. During the startup phase the Investor and Builder classes exercise greater control. As patron membership thresholds are met, majority control transfers to the patron membership classes.

Member's Media is still in the early stages of platform development. Thus far, narrative support initiatives have been piloted through existing platforms and services. There is a need for applications and platform functions specifically designed for the demands of the independent narrative community. We are currently fundraising in order to complete the build of the phase-one platform.

Member's Media is proving the power of cooperative practices in the creation of independent narrative media. With more resources we will also show the power of cooperative audience engagement. Our goal in the coming years is to host a vibrant narrative media community with members worldwide and to provide a home to stellar independent narrative work.

Project Name: TimesFree
Completed by: Francis Jervis (CEO, founder)
Location: San Francisco, California
URL: timesfree.co

TimesFree is a platform for swapping services between trusted friends without using cash.

Babysitting co-ops swap sits between families using a simple token system. They've been tried and tested for over fifty years. Sharers save over $1,000 a year compared to families who hire casual sitters. But running a co-op with spreadsheets and an email list takes up too much time. As well as taking care of all that administration, TimesFree will offer comprehensive safety coverage. Members will be protected by identity verification, background checks and insurance. Just fixing babysitting will make life better for millions of families, but co-ops offer a perfect model for sharing everyday services like dog walking and other errands too.

TimesFree is currently privately held and will be working toward benefit corporation status to help us continue to both serve our members as effectively as possible, and make safe and efficient cooperative, cash-free sharing available to everyone.

So far, the company has been bootstrapped. I built the iOS app (released in August 2015) in Swift, with a MongoDB-hosted back end. A version for the Web and Android are next.

Delivering a user experience that's as good or better than "sharing economy" platforms like Airbnb, and handling user data—especially identity and reputation information, once we start offering services like background checks—are my biggest priorities for developing the next iterations of the service. The "sharing economy" so far has figured out how to make cooperation easy and safe, and it's time to build platforms for real sharing on that scale or bigger. We're only beginning to see the possibilities for real, money-free sharing platforms, and I'm still learning how creative people can be in what they want to share!

Project Name: Snowdrift.coop
Completed by: Aaron Wolf
Location: Earth (incorporated in Michigan)
URL: snowdrift.coop

This book took time to write, edit, and promote. To fund such work, publishers use legal and technical restrictions to make access exclusive to those who pay. Funding is necessary, but restrictions have terrible side-effects, including blocking sharing, discouraging derivative work, and excluding people—ultimately, limiting the work's value. Snowdrift.coop is developing a platform to fund creative projects without artificial restrictions such as those listed above. Our matching pledge creates a network effect: each patron's monthly donation to their favorite projects is based on others joining them, such as a pledge of $1 for every 1,000 patrons. This flexible approach minimizes risk and maximizes collective impact.

As a multi-stakeholder co-op, we propose three member classes: the worker class made of employees of the platform itself; the project class made up of those funding their creative work; the general class made up of users who only donate. As the products are public goods, the only exclusive value of co-op membership will be in decision-making. Each class will have Board representatives, and policy votes put to members will require approval from all classes. As a nonprofit, we have no stock; to get co-op membership, a user pledges to the platform itself as a project.

Everything we do aligns with cooperative values including using exclusively free/libre/open resources. We use the Haskell-based Yesod web framework, create illustrations with Inkscape, and communicate with Jitsi Meet and IRC. We're an all-volunteer organization aside from our web development contractor. We ran a fund-drive in 2014 and continue accepting donations toward our launch.

Since our first public announcements in 2013, we've attracted hundreds of test users and dozens of volunteers, but we face the same challenges as other projects we aim to support. Our published writings and research are interesting, but the real value will come with reaching a working beta stage.

Project Name: Resonate
Completed by: Peter Harris
Location: Berlin, Germany
URL: resonate.is

Resonate is building a streaming music platform with a truly unique listening model—"stream to own"—which helps casual listeners become dedicated fans. This platform in turn allows us to do something that no other service can claim to do: pay musicians directly for every single stream. The first stream of a particular song starts out really cheap, while repeat plays gradually increase in price until reaching the normal price of a regular download.

Resonate is a multi-stakeholder cooperative where musicians, fans, and staff share in profits and governance roles. For voting procedures, it's a one-member-one-vote structure, while profits are distributed according to the value generated by participants—various values being the amount of streams among musicians, the expenditures of listeners, and the time commitments of staff and volunteers.

Development has started in a number of areas: design, market research, and technology. While significant progress has been made, we won't be able to meaningfully sink into development until we succeed in raising funds. Given the typical investor problem for all platform co-ops, we're going to seek our initial capital through crowdfunding—in particular, by reaching out through all the musicians and indie labels in our network. We plan to make the entire campaign reflect our values by recruiting musicians and listeners to fully participate in getting the word out, and by rewarding volunteers through a points system that may be redeemed in the future for streams or exclusive content.

We're very proud of what has been accomplished so far. Design and branding have been firmly established (while the site continues to evolve), and numerous content items have been written and shared socially as our Twitter followers and newsletter subscriptions continue to grow. Additionally, significant connections have been made with multipliers writing on streaming royalty issues: hundreds of indie musicians and a few labels have joined, and we have recruited numerous staff and volunteers eager to dive into development!

Project Name: Loconomics Cooperative
Completed by: Joshua Danielson
Location: San Francisco, California
URL: loconomics.com

Loconomics Cooperative is an on-demand web and mobile app structured as a platform cooperative where the owners are local service professionals who use technology to connect to a community marketplace to grow their businesses.

Owners pay a $30 monthly user fee that funds the business team who market and operate the platform. Business team employees are also owners and will eventually be paid market-rate salaries. Executives will have their salaries capped at 3.5 times the median income of San Francisco. This ensures that income is used to further develop and market the platform in the best interests of the cooperative. Loconomics will aim to actively counteract the tendency for power to be concentrated at the top by creating an equal opportunity for owners to participate in governance, empowering all owners to influence the activities and choices of the organization, and integrating the wisdom, needs, and ideas of a broad spectrum of its owners into the cooperative.

We've bootstrapped the money needed to create the initial app with reliance from a number of professionals who, along with the founders, will be paid with loan payments over the next ten years. There are no equity investors, and the owner-user fees will be able to cover these payments easily when we reach a couple of thousand owners.

We're just finishing beta testing the platform with a small group of owners before an official launch in May 2016. We had one event in November with about twenty-five local service professionals who were very excited about shaping the future of the platform. Most felt that a $30 monthly fee to access the platform's benefits is a small price to pay. We have a database of a few hundred potential local service professional owners and just as many potential clients.

Project Name: NYC Real Estate Investment Cooperative
Completed by: members of the NYC Real Estate Investment Cooperative
Location: New York City
URL: nycreic.com

The New York City Real Estate Investment Cooperative (NYC REIC) is a democratic financial organization that exists to secure permanently affordable space for civic, small business, and cultural use.

Consistent with the principles and spirit of the cooperative movement, the NYC REIC aims to make long-term, stabilizing, and transformative investments for the benefit of our member-owners and our communities. We will: assist communities in raising the capital they need; work with community-based organizations to plan and implement their real estate development projects; and support local community activism to ensure that the city emphasizes affordable, community-controlled commercial space in its land use decisions.

While member investments are the heart of the cooperative, during the startup phase of the NYC REIC, charitable contributions support operating expenses. Once we have identified a few potential investment projects and properties, we are committed to engaging with residents and community-based organizations from the neighborhoods where those properties are located.

We have been working together since May 2015. In just one year, we have organized 350 members into seven active working groups that have met over 150 times, raised $1.3 million in investment pledges, and democratically elected our first steering committee. We have support from 596 Acres, Fourth Arts Block, Spaceworks, Brooklyn Law School, and Fordham School of Law, and are in touch with groups that have been inspired by us in six cities nationally. Our public meetings have over seven hundred RSVPs, and we regularly reach capacity. We know that this is at least a ten-year project. We know that we cannot have the city we want without informed, active residents. By building a cooperative, we are educating, empowering, and shaping a powerful group of New Yorkers who say: development without displacement is possible.

Project Name: Robin Hood Collective
Completed by: Akseli Virtanen
Location: Helsinki, Finland, and San Francisco, California
URL: robinhoodcoop.org

We offer distributed capital services, personal hedge funds, next-generation option structures, and a range of powerful finance tools for everyone. Our motto: the king's options for all. We think that finance is a place of creation. More precisely, we build financial tools for the socially networked generation—for new economic agents, doers, makers, co-creators, and peers, who are also often collective, and don't have access to financial tools or knowledge. We also explore the political and organizational potential of finance, derivatives, and securitization.

We have different forms: a cooperative that is owned by its members (Robin Hood Asset Management Cooperative); a startup owned by the core team and its partners, who have committed to its special purpose of building a next-generation finance platform (Robin Hood Services); and an open-source platform for distributed financial tools, owned by all of its users (Robin Hood Unlimited). This is a dynamic organizational formation that is never exhausted in its actual forms. You can't own a multitude.

We started with the cooperative but understood that it was an organizational form that belonged to the last century, to an industrial understanding of sharing of risks and rewards. We needed a more dynamic, more multidimensional, more distributed way of owning, financing, doing things together, and risking together. More equity, more options, more assemblage, more king's deer for all—that is what we are building now. Blockchain technology offers the perfect organizational infrastructure for that.

We will launch the world's first hedge fund on the blockchain this spring, and the next new financial services—DotCom Mutual Unicorn, HouseHold Union. and DistributedCapital—later in 2016. What makes us most proud? What we do touches people's imagination.

Project Name: Seed.Coop
Completed by: Rylan Peery
Location: Ithaca, New York
URL: seed.coop

Seed.Coop helps organizations grow by providing tools and support services for easy onboarding of new members.

Seed.Coop is currently held in trust by CoLab Cooperative and will be structured in a way that meaningfully gives control and ownership to a diverse set of cooperative members and stakeholders (see http://bit.ly/colab_in_trust for a full explanation).

The prototype of the platform has been built via sweat equity by core team members. CoLab Coop has provided sponsorship funds to underwrite a percentage of the development. The steering team is currently exploring outside funding sources and pilot organizations to accelerate development of the platform. Of particular importance is cooperative support among organizations so that there are no silos between orgs building membership. Instead, members share a symbiotic relationship via a member "referral engine."

We have built our first prototype of the platform and are now working on prototyping decentralized membership onboarding solutions for specific partners. We expect that, in 2017, Seed.Coop will be a cooperatively owned platform for member onboarding working interoperably with platforms like Coopify and The Working World.

14. THE REALISM OF COOPERATIVISM

YOCHAI BENKLER

Cooperativism, or mutualism, has been in the repertoire of alternatives to capitalism since nineteenth-century figures like Owen and Proudhon. In some regions—the Basque Country, or Emilia-Romagna in Italy—or industries—U.S. dairy farming, for instance—cooperatives have become major, sustainable parts of the region or sector. But we have to be honest: cooperativism has not played a transformational role in the past two centuries of capitalism.

Four dimensions of opportunity suggest that the future could be different.

First, disruption. Things are very much up in the air. Uber is growing dominant in personal-transportation services in the United States, but Uber could still be the Friendster, or at most the LinkedIn, of the on-demand economy if the cooperative movement moves fast into a broad range of services. Historically, cooperatives have been stable in the face of market competition where they did emerge, but not sufficiently competitive to force their way into markets already saturated by conventional firms. Conventions, imitation, and practice—not economic superiority—determined the presence or absence of cooperatives. The moment of opportunity is now, when the organization of production is still in flux.

Second, we are in the cultural moment of cooperation. Wikipedia, free and open-source software (FOSS), citizen journalism, and other forms of commons-based peer production have made normal people encounter cooperation and its products as a matter of everyday practice. The decades-long insistence of expert economics that we should think of ourselves as self-interested rational actors acting with guile is bumping up against a daily reality that refutes it. Sciences, from evolutionary biology through the social sciences, psychology, and neuroscience, are all lining up to confirm that people are not the moral midgets and sociopaths that populate game theory and rational actor modeling—that many of us cooperate when we are in situations we understand as cooperative, and compete when we are in situations that we feel are competitive.

Practice and theory are providing the cultural framework within which people can come to believe that cooperativism can in fact work, on a mass scale, for important swaths of their Internet-mediated social practice.

Third, commons-based peer production has provided a template and experience with the possibility of large-scale enterprises managing and governing themselves through online cooperative platforms. They offer extensive and growing experience with how networked peers govern themselves, allocate work and responsibility, and manage day-to-day operations across time and space. Peer cooperativism shares these core governance and organizational patterns with commons-based peer production, but its defining feature, enabling workers to make a living from their cooperative work, presents distinct challenges that peer production has not had to face.

Finally, networks have destabilized the model of the firm. Transaction costs associated with both market exchanges and social sharing have declined; interactions once preserved for firms that combined capital with contractual commitments for labor, materials, and distribution can now be done in a more distributed form. This technological fact has underwritten the rise of the on-demand economy, workforce management software that increases contingency, and outsourcing and offshoring no less than it underwrote FOSS, Wikipedia, or SETI@home. It will not determine a more cooperative future, but it does mitigate some of the most important barriers that historically hampered cooperativism. Uber and Airbnb both involve the reallocation, through markets, of shareable goods: mid-grained lumpy goods, put in service by people for their own use, but with excess capacity. When those who enthuse about these platforms emphasize that they have built the largest transportation or hospitality company without building a single room or owning a single vehicle, they are meanwhile pointing out that the barrier to effective scale for cooperatives—the need to invest concentrated capital—is less strictly constraining than it once was. Commons-based peer production has shown us that software can be developed as FOSS, marketing can be done peer-to-peer, and it is likely that the legal and political meaning and contests over services will be fundamentally more legitimate and less oppositional when the service is built by cooperating peers.

The combination of economic disruption and the opportunities to capture new markets, a shared cultural imagination of the possibilities of cooperation, and deep practical experience with online cooperation as a practical solution space make this moment different than it might have been two decades ago, much less in the heyday of industrial capitalism. Maybe.

There are real challenges before peer cooperativism can occupy a substantial space in the networked economy. Peer production has thrived on pooling voluntary contributions of participants who had other means of making a living. This allowed commons-based peer production to release its outputs mostly free of charge, as well as "free as in freedom." Peer cooperativism, if it is to become part of the solution to the increased economic insecurity for the many in the twenty-first century, must be able to sustain cooperation while charging customers and users a price and fairly distributing the proceeds among the peers. This is a challenge that commons-based peer production did not face. The established cooperative movement has shown that the challenge is not insurmountable, but it is real. Not least among these challenges will be the need to mediate the driving ethic of peer production, ensuring that its outputs are in the commons and available for all, with the necessity of providing income to the peers themselves. This will be easier for service models, as we have seen with FOSS, than for information goods that do not have a clear service model, like stock photography. Ethical coherence strongly suggests that cooperatives providing information goods must develop models of shared membership or service, rather than aim for building on an "intellectual property" strategy that will separate these cooperatives from the heart of the movement.

Cooperativism is not simply shared ownership, as are many employee-ownership plans. It is, first and foremost, shared governance. Oscar Wilde is supposed to have said "the trouble with socialism is that it cuts into your evenings"; the labor involved in peer cooperativism presents a challenge, although online democratic governance platforms offer flexibility, transparency, tracking, and discourse-flow affordances that can make the load more manageable than it was for physical world cooperatives. Some FOSS projects, like Debian, have successfully developed a democratic process. Many others depend on a charismatic-leadership model that may not translate well into the domain of making a livelihood. The primary resource for platform cooperativists must be the rich literature on commons governance pioneered by Elinor Ostrom, which did in fact focus on collaborative communities managing their livelihood resources together without property rights or government laws. The growing work on governing knowledge commons is the best source of translating between the Ostrom school and the experiences of online cooperation, albeit without the focus on making money and distributing it.

The enormous literature on governance in Wikipedia will be pertinent, for instance, because Wikipedia, unlike many other peer production communities, has evolved into a body that has a responsibility—cultural, if not economic—for an output. And Wikipedia tells us that things won't be easy. Combining lessons from the rich literature on Wikipedia governance with the Ostrom-school literature must drive cooperatives to design not only participation, but also mutual monitoring and dispute resolution systems, and in particular affordances to permit nested power or subsidiarity—the organization of governance at the closest possible level to where the activity being governed is taking place, while coordinating across the cooperative. The biggest likely divergence from peer production will be the need to define membership more strictly. In cooperativism, as with commons-property-regimes, it will be important to clearly define who members are, and place a higher barrier on membership than peer production has done. This is so partly because the quality and timing of outputs will be more critical, and partly because of the need to maintain a reasonably defined universe of participants among whom returns sufficiently high to make a real contribution to their livelihood must be shared.

Decades of studies of cooperation tell us that communication among members—particularly communication that humanizes members to each other—is central, as is developing a shared identity. A strong core of moral values, avoidance of an ethic of "I'm just here for the extra few bucks," and a clear commitment to fairness among the members will be necessary to overcome the inevitable tensions associated with work and income sharing. Framing is important, and while self-interest undoubtedly plays a role in any community, no effort that appeals primarily to that self-interest will likely survive, let alone outperform explicitly self-interested models of investor-owned firms.

At no time in the two centuries since cooperativism first appeared as a conscious alternative model to modern organization of production has it been more feasible. That it is feasible, however, does not make it inevitable. As a movement, cooperativism will only succeed by moving fast and decisively, learning from the near past, and sharing our experiments and knowledge quickly and repeatedly in a network of cooperatives.

15. THREE ESSENTIAL BUILDING BLOCKS FOR YOUR PLATFORM COOPERATIVE

JANELLE ORSI

Below, I describe three elements I believe should be encoded securely into the legal structures of platform cooperatives. These are particularly designed for cooperatives that aim to be of long-term benefit to their members.

Oh, wait. Sorry about my redundant sentence. I basically just said that these elements are for cooperatives that aim to be *cooperatives*. That *is* the purpose of cooperatives: to benefit their members.

I had to start with that reminder because I think it's hard for most of us, having embedded ourselves into the dominant models of doing business, to break away from certain notions about what a tech startup should do and how it should work. It's enormously challenging to narrow in, with unclouded and unwavering focus, on just one thing: to be of benefit to a community of platform users.

I frequently hear people talk about cooperatives as if they are a plugin that can be installed into a run-of-the mill corporate structure. But cooperatives are a completely different operating system. They process things quite differently, and their outputs are different. In a world where the dominant models of doing business are widely recognized to be escalating inequality and destroying the planet, we desperately need to build economic operating systems that achieve the exact opposite. Cooperatives can be such an operating system if we build them with great care, adapt them constantly, and install every possible protection from infection by business-as-usual.

When I started practicing law eight years ago, I committed to focusing my work on supporting cooperatives. I've now seen both successes and failures among cooperatives, and I'm perpetually challenged by questions like what makes cooperatives work and endure, and how can they manifest true economic democracy? Of course, these questions merit multiple volumes. For now, I'll share three things that I would advise any platform cooperative to build into its legal structure.

Before I dive in, though, here are a few concepts that guide my thinking on platform cooperatives:

First, platforms are us: Platforms aren't just software applications and the companies that administer them. What gives a platform value, in most cases, is the community of users that employ the platform, along with the networks, data, and ideas they create. In other words, what makes platforms so valuable is what we put into them.

Second, platforms don't need to be treated as commodities: It's easy to develop a platform fetish as a result of their seemingly magical potential to create billionaires. Yet all along, it is the users themselves, and the rents they pay to platform companies, that enable the billion-dollar valuations. Cooperatives, however, exist to provide benefit to their members, and they do so by *not* charging rents to begin with. It would make no sense for a cooperative to charge excessive fees to users—it would just end up paying the fees back to the same users in dividends later.

When the spell of the platform fetish has been broken, we can go back to focusing on the primordial function of platforms: benefiting users. Maintaining this focus won't be easy—not when we are pulled back into business-as-usual thought patterns nearly every time we interact in the world. This will get easier the more that we come to be surrounded by other cooperatives. Less and less, we'll find ourselves asking things like: How can we market to a wealthier class? How can we raise prices? More and more, we'll ask things like: What else can we do to help our users? How can we lower prices? It sounds strange at first, but transitions like this can make us wake up every morning feeling slightly more grounded and human than the day before.

In the meantime, we can install strong protections that defend our cooperatives from even our own habitual ways of thinking. Fortunately, cooperatives come with at least two pre-installed protections, which are built into cooperative legal structures: First, money doesn't buy power in a cooperative; cooperative members elect the Board of Directors on a one-member one-vote basis. Second, money doesn't buy profits in a cooperative; cooperative members may receive dividends, but the dividends are calculated in direct proportion to money spent by or earned by each member, not on the basis of how much money anyone had the privilege to invest. These are powerful protections in a world where money seems to buy power and profits everywhere else.

Beyond those pre-installed protections, I'd urge that a few other principles be embedded into cooperative bylaws and policies. Here are my three big building blocks:

1. Prevent the platform from being sold. Imagine that your platform cooperative builds a vibrant community of ten thousand dedicated member-users. Then a large company or investor comes along and offers your cooperative $50 million to take over ownership. Tempting, right? Each member could get a payout averaging $5,000. But the potential for quick cash can blind people to their long-term self-interest. A for-profit company could start charging higher fees to users, selling their data, manipulating search algorithms to privilege some users over others, and other practices that put users at a permanent disadvantage in relation to the company.

It is critical to adopt strong safeguards against the potential for the platform and any of its major assets to be sold into for-profit structures. A decision to sell major assets of the platform should require a high threshold of approval, such as by 80 percent of members, after what would hopefully be a long public conversation about alternatives. I would even suggest that outside parties, such as a panel of nonprofits and other cooperatives, be given the opportunity to review any proposed sale, veto it, or exercise a right of first refusal to buy the assets. Lastly, to remove a major incentive to sell to a company, I recommend capping the amount of sale proceeds that can go to members and sending the rest to a nonprofit.

2. Put a cap on pay-outs and compensation. In our competition-driven economy, a scarcity mentality has given many people—or even most—an insatiable drive to accumulate wealth. Even a cooperative can get swept up in this dynamic if powerful stakeholders use their leverage to extract value from the cooperative. Executives could vie for higher and higher pay. Although cooperatives generally—and preferably—do not allow for profit-maximizing equity investment, a cooperative could still end up giving up too much to lenders or preferred shareholders. To prevent these possibilities from even making it to the negotiation table, a cooperative's bylaws should establish caps on employee pay, investment return, and other payouts.

Where to set those caps? The platform cooperative Loconomics uses data from the Bureau of Labor Statistics to cap employee pay at 3.5 times the median wage for all occupations in the region where the employee works. In the Bay Area, the cap would be $165,000. Among tech CEOs, that might sound like poverty. But to an average person, that might sound like, well, a whole 3.5 times better than life as it is. It's all relative, and the notion of "enough" is completely lost in a society with such dramatic income inequality.

When decisions are no longer driven by the desire to maximize gain, I think that a desire to ensure that everyone has enough is the ethic that steps in to replace it. True, a company won't be able to attract the kind of talent that thinks of everything in terms of commodities and maximizing monetary profit. But it will attract talented people who are smart enough and have enough self-awareness to know that doing good work with good people is what makes for a good life. If you hire someone who is driven not by profit-maximization, but rather by a desire to do meaningful work, they will also be more intuitively oriented to the cooperative's purpose of benefiting members. Which leads me to the third item...

3. Adopt a staff trusteeship model of governance. Staff trusteeship is a governance model that views all staff members of the cooperative as trustees who manage the platform for its beneficiaries, the body of members as a whole. I believe that staff trusteeship is a natural model for large cooperatives in particular. Democracy in larger cooperatives is often quite bare-bones, mostly limited to members voting for directors. Members' voices are easily lost if they are primarily expressed through the election of a board that meets infrequently and is not very tuned in to the day-to-day work.

In a staff trusteeship cooperative, every staff person becomes a point of accountability for the organization, taking on responsibility to listen to and amplify the voices of its members. Staff self-manage using an internal governance and operational model such as Holacracy or sociocracy, which distribute power among staff, removing inefficient hierarchies and ensuring a great deal of agency for each staff member.

In a staff trusteeship cooperative, the board of directors is still elected, but it takes on a role that is more akin to that of a guardian, overseeing the activities of the organization and ensuring that the staff are tuning in to members in every possible way. Incidentally, staff will generally find their work far more interesting and rewarding in this model. They are not there to merely execute the directives of a board or CEO; they get to bring their full selves to work, to apply their talents, engage their full potential, and work with members to solve real-life problems.

The list of things I would build into cooperative bylaws is actually much longer than the above, but I have emphasized just three here, because I find them particularly useful in jolting us out of business-as-usual ways of thinking, and in getting us back to the pure focus on being humans working to benefit humans.

By the way, you can replace the word "platform" in this article with "housing," "land," "workplace," "water," "food," or "energy," and apply the same principles to any cooperative. The beauty of platforms, however, is that they are far easier to cooperativize than, say, land and housing. Platforms are us. We don't need to mobilize enormous amounts of capital to build a cooperative economy in the world of tech. We primarily need to mobilize ourselves to make different choices. Then, having done that, we will have built collective power, which gives us a platform of a different sort: one on which to launch cooperatives of all varieties to take back land, housing, water, energy, jobs, food, and everything else that has been poorly served by business-as-usual.

16. SO YOU WANT TO START A PLATFORM COOPERATIVE...

CAROLINE WOOLARD

Dear Founder,

I'm glad to hear about your idea for a cooperative platform. Congratulations! I'm sure we both agree that a diversity of opinions is a good thing, and that platforms should benefit their participants, as participation is what makes a platform valuable. What follows are a few questions that I wish someone had asked me when I started four multi-year projects, most of which continue to run today.

The projects I co-founded, for what it's worth, are an 8,000-square-foot affordable studio space (Splinters and Logs LLC, 2008–2016), a resource-sharing network (OurGoods.org, 2016–present), an international learning platform that runs on barter (TradeSchool.coop, 2010–present), and an advocacy group for cultural equity (BFAMFAPhD.com, 2014–present). I also helped convene the first gatherings of the NYC Real Estate Investment Cooperative in 2015 with Risa Shoup and Paula Segal, and am inspired by the ongoing work of the NYC REIC's member-elected steering committee and the open working groups.

I am sharing these four questions, along with bits of advice, because I hope that you will succeed in contributing toward the cooperative culture we want to see. To live in a democratic society, we all need more experiences of democracy at work, in school, and at home. Thank you for helping push the cooperative movement forward.

You will notice that a lot of what follows also speaks to founders of nonprofit organizations or social impact businesses. I am writing this especially for young, educationally privileged people who have big ideas but are newcomers to the neighborhood they live in. This reflects my own experience as a college graduate, waking up to working class histories in New York City while trying to build cooperative software and resource-sharing projects. It took me a while to learn outside my immediate group of friends; to reach beyond the academy and beyond the Internet to learn.

1. Can you make a platform for an existing co-op?

In a culture that values ideas over practices, it might be hard to see the existing cooperatives around you. But, I promise you, there are many systems of mutual aid and cooperation nearby. These "platforms" are systems of self-determination and survival created by people who have been systematically denied resources through institutionalized racism, sexism, and classism (read about redlining if you don't know what that is). The credit unions, land trusts, worker-owned businesses, rotating lending clubs (susus), community gardens, and freedom schools in your neighborhood may not have great websites, but they are incredible cooperative platforms that you can learn from and with.

These initiatives are often not lifestyle choices made by educationally privileged people, and will therefore not be written up in *The New York Times*, but they are robust and powerful community networks with organizers who *might* be interested in adding an online platform to their work. Here is an often-overlooked challenge: try to join and add to existing cooperative platforms, rather than building your own from scratch. The result will likely last longer as it will be informed by the deep wisdom of existing cooperative community norms, roles, and rules. Perhaps we need something like the Center for Urban Pedagogy for cooperative software—an organization that matches grassroots groups with developers to build software that is driven by community need.

2. Who will build the cooperative platform?

Let's say that organizers at your local credit union, land trust, cooperative developer, community garden, or freedom school are interested in building an online cooperative platform to add to their ongoing work. Or, they confirm your hunch that the cooperative platform you want to build is necessary. How will you form a team that can make this software come to life?

I have found that innovation occurs most readily in small teams with shared goals but different skill sets. Big groups, on the other hand, are good for education and organizing work, and for refining existing platforms. But to innovate, I like to work in core teams of three to six people, as this allows for deep relationships, shared memory, and relatively fast decision making, since each person can speak for ten to twenty minutes per hour in meetings. The collective Temporary Services says that every person you add to the group doubles the amount of time it takes to make a decision. So, I say: build a small group of rigorous, generous experts whose past work demonstrates that they are aligned with the cooperative platform you want to make. Ask the larger group to consent to the expertise of your small team, and ensure that your small team will make room for feedback from the big group along the way.

Now, build your team! Find people who are better than you in their area of expertise. At the very least, you will need: 1) a Project Manager to help with scheduling events, facilitating meetings, and tracking budgets; 2) a Communications Pro to craft a clear message and recruit people to try out the platform as it develops; 3) a Designer (or two) who makes the front end beautiful, 4) a Developer (or two) who develops the software and annotates it so that other people can add to it in the future; and 5) Advisors—one per area of expertise above, as well as more who have strong connections to the community you aim to work with. Meet with your core team on a weekly, if not daily basis, and with your advisors on a monthly or quarterly basis.

You are likely the Communications Pro or the Project Manager, since you are reading this letter. Find advisors who are retired, or far older than you, and who have seen the field change and are widely respected for their work. Learn about programming languages—which languages (Ruby, Python, etc.) have active development communities, and which languages are most likely to be interoperable with future cooperative platforms. Find developers who have worked on social justice projects in the past. If you are a nonprofit with limited funds, watch out for developers who want to get paid market rate, as developers and project managers (like you) should believe in the project equally and should take an equal pay cut. Watch out for developers who say they can build the site in a public hackathon or sprint, because if they do that it won't be built well.

3. How much time and money do you have?

As you build your team, be honest with yourself about your existing priorities, and the likelihood that your life will change in the coming months or year or two. To gauge our availability to work on TradeSchool.coop, we did an exercise where each core member wrote a list of their top life priorities, including family, friends, health, volunteer projects, art, hobbies, and day jobs. This allowed us to be more honest with ourselves and each other about the amount of time we had to work on our project, which parts of our life were unknown, and also our reasons for doing the project.

Plan for turnover by having clear systems of documentation and open conversations about how to bring in people who might join the core team when someone has to leave. Be sure that the Developer(s) code in teams, or that an Advisor looks over the code, so that it is intelligible to your other Developers. Be sure that the Project Manager and Communications Pro share leadership and responsibility, crafting a clear process for new people to join the core team, moving from roles of assistance to core membership in months. After a year of organizing TradeSchool.coop, I wrote a manual to make sure our systems were clear. Ask yourself: do you want to get it done, or do you want to get it done *your way*? This is the question that Jen Abrams, a co-founder of OurGoods.org, brought to us from a decade at the collectively run performance space WOW Café Theater.

4. What if you ran events and hired a community organizer instead of building software?

Last of all, consider the possibility that you could make a greater impact on cooperative culture and resource-sharing in your community by hosting events rather than building a new cooperative platform online. Software does not run itself; it must be maintained and upgraded by developers who can easily make tons of money working on non-cooperative platforms.

Remember that people won't take the time to learn a new app unless they need it daily. Remember that people are used to Facebook, Google, Twitter, and sites that have legions of developers working around the clock. Remember that hire number three at Airbnb was a lobbyist. If you are starting out, build the smallest feature and do not add to it. It will be hard enough to maintain and upgrade that small feature.

Be honest about your ability to put in long hours and to raise the funds to sustain the development and constant upgrading of online networks for years. Until we have cooperative investment platforms for cooperative ventures, you will have to look for philanthropic support or venture capital that might alter your mission and that will rarely sustain the initiative for years.

If you can't raise $300,000 a year for a core team of five, don't build a demo site that barely works or buggy software that won't last—organize great events and build community! You can use existing online platforms that your members already know. You can use your funds to pay a community organizer instead. Not only will you sustain the livelihood of a wonderful person, but the knowledge built in the community won't return a 404 Server Error when someone needs help next year.

In cooperation,

Caroline Woolard

PS: If you want more information, just email me at carolinewoolard @gmail.com. I also put a lot of links to organizing, facilitating, and horizontal structures in the *How to Start a Trade School* manual from 2012, and the NYC Real Estate Investment Cooperative's REIC U working group is making a long list as well. Look for it on NYCREIC.org!

17. WHAT WE MEAN WHEN WE SAY "COOPERATIVE"

MELISSA HOOVER

Is platform cooperativism a social movement or a market intervention? I would argue it is both, but that we should clarify the distinct implications of each of those imperatives—how they could come into conflict with one another, and where they can intersect powerfully.

These questions—these tensions, and their power—are not unique to platform cooperatives. They animate the cooperative form at its core, as cooperatives sit squarely at the intersection of values and markets, organizing and business, community institution and economic engine. Understanding this nexus with as much clarity as possible will help us build successful platform cooperatives and the appropriate ecosystems of support for them.

As we advance our thinking on platform cooperatives, there seems to be some impulse to muddy the concept of cooperatives, even to dilute it so thoroughly that it comes to mean "something we like," or "something based on values," or "something not owned by outside shareholders." All of these are partially true, but none captures the promise of the form completely. It is critically important not to dilute the meaning of "cooperative" as we advocate for it on a broader scale. Rather, we should let the concept's clarity move us toward simpler and clearer formulations of what we mean when we say "platform cooperativism."

At their most basic, cooperatives are values-based businesses that operate for member benefit, that are owned and controlled by the people who do business within them. They are formed to meet those members' needs, and the nature of the members' economic relationship to the cooperative determines what kind of cooperative it is. Though centered on member and community benefit, cooperatives are not nonprofits; they operate in the market and are subject to market forces, although they often arise where conventional markets have failed to meet people's needs.

Carrots can help us understand some of the types of cooperatives:

- If the members' need is to buy carrots, they may form a cooperative that buys carrots in bulk and sells them to members. This is a **consumer cooperative**: the members are consumers, and the cooperative helps them access products at a fair price.
- If the members are carrot farmers and their need is to sell their carrots, they may form a cooperative that pools their carrots, sells them under a shared brand, and gets the best price they can in the market. This is a **producer (or marketing) cooperative**: the members are independent producers and the cooperative helps them access markets to sell products at a fair price.
- If the members' need is for paid work, they may form a carrot processing plant that buys carrots, adds value through their labor, sells the carrots, and uses the income to pay members. This is a **worker cooperative**: the members are workers and the cooperative helps them access good jobs.
- If a cooperative is designed to meet multiple types of needs, it may have multiple types of members. This is a **multistakeholder cooperative**: carrot growers connect to carrot distributors and carrot consumers in one holistic entity that aggregates all three pools of membership.

Amazon, of course, isn't selling carrots; it's selling convenience and logistics. Uber isn't a taxi company; it's a lobbyist, a loan shark, a labor broker. Developing platform cooperativism past its infancy stage—or more to the point, intervening in platform capitalism as it reaches the terrible twos—requires a keen ability to discern what is really being sold, how the money actually flows, and who benefits. A return to the fundamental questions that structure cooperatives—who are the members, what is their relationship to the cooperative, and how is it meeting their needs?—can sharpen our analysis, bring much-needed clarity to a complex concept, and help identify effective interventions that center on worker and community benefit.

Workers' needs clearly are not being met by current platforms. Platform capitalism removes any accountable mediator between capital and labor: there is no management to petition, no corporate structure to organize against, just the platform with its built-in discipline of user ratings and a contingent labor pool fathoms deep. Meet the new boss, same as the old boss—except without, you know, any actual boss, just the unmitigated imperative of capital to return value to investors. Cooperatives actually connect investors directly to markets, too, but in a very different way: the investors are members of the cooperative itself. This alignment of interests can capture the promise of the platform—direct connection to distributed markets—while centering worker benefit as its reason for being.

In the context of the platform economy, the old distinctions among types of cooperatives still matter immensely. Why? The relationship between owners and employers remains at the heart of many platform capitalism models, although the platform owners attempt to turn our attention away from it. Cooperative forms inherit this challenge; they don't automatically solve it. Neither platforms nor cooperatives are so revolutionary that they obviate the need to address workers' rights and protections. A platform—even when owned cooperatively—is still simply brokering market access and labor relations. For these arrangements to be fair, they need to be clear.

Platform co-op developers should think carefully about what kind of cooperative they intend to create. For instance, there may be good reasons to set up a platform cooperative of producers, rather than one based in an employment relationship. Taxi drivers may want to retain autonomy over their earnings and flexibility of gigs; house cleaners may want to maintain a direct economic relationship with a client; both may want a platform to increase their access to a market. Owning and controlling the platform as a cooperative of independent operators ensures that it serves them, rather than extracting from them. For consumers, producer platform cooperatives may offer greater variety or availability of service providers, while uniting them under a common brand that conveys trust or increases access.

Similarly, there are reasons to set up a platform worker cooperative. Care workers and their clients may want the protections that come from a client-facing employer entity; owning and controlling the platform as an employment entity provides both the protections of employment and the workers' right to set the terms for a market they rely on. For consumers, the accountability and reassurance offered by an employer entity may be critical; people may be reluctant to entrust their aging parents' in-home care to a stranger from the Internet, and only an institutional relationship can appropriately mitigate that risk.

In any of these cases, membership is meaningful, equating to ownership and control over the entity. A platform that doesn't actually consider at a granular level the question of membership, its members' needs, and their relationship to the cooperative—one that uses "cooperative" as some sort of trust mark not backed by actual cooperative structures—runs the risk of simply being part of the problem. Marketing the brand without the structure could end up replicating and reinforcing contingency and, more practically, resting on a very shaky foundation as a business.

Millions of dollars of venture capital are pouring into platforms designed to exploit our desire for convenience while destabilizing entire workforces—and often still not achieving market viability. The advantages of this platform capitalism, meanwhile, are being aggressively marketed and consolidated. We must therefore be especially clear about the added value of the "cooperative" in platform cooperativism. Platform cooperatives will be worthy of consumer trust only because they are structurally built on trust: members operating a business for member and community benefit.

In the platform context, the added value of cooperatives comes from the democratic commitments that they operate around—and clarity of purpose is necessary to build structures that operate in service of those commitments. As we react to rapidly changing economic and social structures, we can best retool our principles and our strategy for this new world by keeping clear focus on some decidedly old-world cooperative fundamentals.

18. A DIFFERENT KIND OF STARTUP IS POSSIBLE

DAVID CARROLL

The standard template for creating a tech company has begun to crack. In 2016, investors are funding fewer companies than before, and many of those that have been funded are announcing layoffs as their exuberant valuations are adjusted to worsening market conditions. For founders, it's becoming increasingly difficult to raise new or subsequent rounds of funding, as investors regain leverage over the entrepreneurs who have put their livelihoods, careers, and emotional well-being on the line to pursue their big ideas.

Two years ago, after working on a sponsored research project with my graduate students at Parsons, we decided to spin off our machine-learning publishing platform into a tech startup. We worked according to the only model we knew, where you sell your equity to investors on a massive bet that you can become their mythical unicorn success story. Like the 90 percent of new companies that you never hear about because they fail, we didn't win the startup lottery. While our business didn't succeed, we learned lessons about starting a technology company. We also noticed ways in which the landscape is already starting to shift as new technology radically transforms and disrupts markets and opportunities, yet again.

Today, I'm a recovering entrepreneur, still reeling from the side effects of losing our bet. But if another opportunity to build another technology platform presented itself tomorrow, we now have an entirely new labor and ownership model to consider, given that the co-op model is increasingly being adapted to technology platforms. In many ways, the platform co-op model is well suited to counteract some of the ownership and sustainability problems intrinsic to venture-backed enterprises that we encountered firsthand. But near-future tech platforms will be built upon rapidly evolving infrastructures and will require sudden adaptations to new capabilities. Given technology to come, what assumptions should be questioned? Platform co-ops need to be designed for tomorrow's marketplace, not today's. Based on lessons learned from starting a tech company, along with my academic research, I anticipate that the following sets of challenges and opportunities will shape the possibilities for entirely new categories of cooperative businesses.

Platform co-ops can benefit from the bursting of the content bubble. The Web suffers from the dominance of cancerous impression-based advertising, and so new business models for producing content are badly needed. The same pressure from walled-garden social sharing platforms like Facebook, Instagram, and Snapchat monopolizing app audiences also threatens the viability of the commons and the Open Web. Its content is largely supplied by the legacy publishing industry to be consumed by conventionally profitable audiences measured by impressions using invasive industrial surveillance technologies. As people adopt ad-blocking software on a linear growth curve past the 200 million users of today, the traditional publisher business model is threatened. An increasing reliance on discovering content through sharing on the big platforms' news feeds ensues. Impression metrics continue to drive the industry to overproduction, leading toward the risk of a content bubble bursting. Platform co-ops like Member's Media and Resonate can seize upon this destabilization as we seek new models for funding and producing media for connected audiences.

App store opportunities are drying up. At first, app stores were a boon to independent newcomers. Now app stores are crowded, and the select few apps that get installed on home screens compete for our daily attention. App makers are increasingly struggling to survive on these over-saturated storefronts. Indeed, proprietary platforms inherently conflict with the philosophy of free, open, and decentralized

114

systems and so many cooperatives might reject these corporate market-places that take a steep cut of proceeds, arbitrarily regulate access, and concentrate platform market-share. But can you make a big enough impact if you're not on these devices? If you need to, how can you earn a coveted home-screen spot or fit into our communication habits? What if you don't need to build an app, but rather should be building a modular service that integrates into the social communications apps that already consume our attention?

Chat is the new interface for apps. What comes after apps? As people spend most of their time in messaging apps and less and less time in specialized app utilities, functionalities are migrating into our conversational interactions. We are seeing how modularized services are poised to replace apps, where the familiar functionality of apps dissolves into the text messaging prompt of our preferred chat service. Amazon expects its customers to verbally chat with its shopping bot Alexa on their Echo home appliance. Facebook has planned to offer publishers access to its Messenger platform to deliver news as a conversation with users. Slack dominates the workplace communications pipeline and pioneered integrations with other services and bots to meet workers where they already are. Text-based labor platforms like Jana suggest that this is possible in the context of platform co-ops. What if you could connect and support a platform co-op through chat bots rather than relying on your potential customers to download and keep using your app or remembering your website?

With artificial intelligence maturing quickly, it will matter more and more who owns it. Leading computer scientists predict a 50 percent chance that software will substantially write itself by 2050. But we're already seeing more machine learning capabilities woven into the fabric of our daily lives. As apps and the Open Web gradually give way to big platforms, we'll see more and more people converse with machines in natural language rather than tapping icons and finagling user-interfaces. It's possible that the bulk of our utilities and content consumption will be further embedded into our messaging apps as bots bump out buttons. What does this mean for platform co-ops who need to capture the attention of people, communicate with them, and transact following the prevailing business logics, and do this ethically? There is an urgent and critical need to build AI platforms that are co-owned and governed by the people who will use them for more and more purposes. These crucial efforts will help alternative AI software mature and develop toward an intelligence that truly represents us, not just the wealthy few who funded the earliest research and development expenditures.

Cooperative online platforms need a free, open, and radically decentralized answer to the cloud. The cloud is expensive and decentralized platforms are only now emerging. Platform owners claim that the Internet is free, but the conflation of free-as-in-*libre* and free-as-in-*gratis* causes confusion. The corporate cloud, really, is just someone else's computer; it is at odds with platform co-op ethics, especially when we realize we're just renting access and computation. However, to deliver the AI-powered features that near-future users will demand, applications will need to draw upon sophisticated industrial-strength AI software services and harness powerful clusters of data-mining server farms. This stuff doesn't come cheap. Free, open, and radically decentralized AI isn't a thing yet, but blockchain-based platforms like Ethereum and Backfeed could offer decentralized alternatives to the corporate cloud. More *libre* but not *gratis*, as you'll pay for decentralization with cryptocurrency. In its infancy, Ethereum is far more expensive than the Amazon cloud but with laughable performance and capability by comparison. Can you afford to wait for the decentralized solution or do you accept that a corporate cloud is presently your only viable high-performance and affordable option?

Co-ops require novel legal frameworks. Starting a conventional tech startup incurs tremendous legal costs that even make the cloud seem like a bargain. Co-ops require complex legal negotiations, which demand specialized legal expertise. With our startup, we sunk some of our common stock equity into our lawyer and accrued additional, deferred, unpaid legal fees to build our entity and negotiate deals.

And by the time we had a thick corporate book of complex investment and intellectual property agreements, we hadn't even gotten to writing up our privacy policy and terms of service, both of which would have been unusually expensive because we didn't want to simply adopt aggressively invasive boilerplates. Without formidable legal prowess, the business world can eat you alive, especially in software. Perhaps the most important early innovation in platform co-ops will be free and open legal frameworks and new attorney compensation models that eschew the conflicts of interest pioneered by Silicon Valley lawyers.

Platform co-ops can mitigate dark surveillance patterns. Can you mitigate surveillance and dark AI patterns? Building data-driven technology can be scary. It's horrifying how easy it is to build a behavior-tracking infrastructure with modern web frameworks. Everything gets much worse when you have to extract payments from people. The software and metrics begin to write their own behavior-tracking algorithms. When AI takes over, particular attention has to be paid to surveillance-related design elements. Unchecked, they could pose grave dangers. Platform co-ops can succeed at building privacy-positivity and basic decency into products and sell this as a competitive advantage against venture capitalist-backed tech companies that lack such qualities because they practice what is increasingly recognized as surveillance capitalism, the extraction of our data to modify our behaviors at scale.

Learn from those who are succeeding already. Stocksy United is winning by being a design-led co-op that serves its design-oriented customers through co-ownership. Loconomics is gaining traction in the micro-labor market by solving pain-points of customers that VC-backed startups don't even touch, such as certifying the safety credentials of service providers. Fairmondo is a platform co-op Amazon-style retailer. But dig deeper to find out who has already attempted your idea and investigate why they failed. The site autopsy.io chronicles failed startup stories. Timing is likely the most determining factor for your success; something that failed before might work this time around as conditions evolve and our expectations shift.

The time is right for platform co-ops. East Coast and West-Coast VCs have predicted 2016 as a year of price and value correction in the tech startup world. Ultimately, extreme decentralization by block-chain may prove ungovernable, at least initially. Governance and shared ownership form the basis of platform co-ops, which are distinct from the C-Corp, LLC, or even B-Corp (Public Benefit Corporation). As tech startup workers begin contending with severe tax penalties now that their stock is underwater, re-priced by institutional investors in down-rounds, the lure of laboring for VC-backed tech startups could begin to wane. As Silicon Valley's supremacy falters, platform co-ops purveying new tech are well poised to offer a better kind of web, one that works more equitably for the people that create and use it because it promotes social justice rather than heralds dystopia as prophesied in science fiction co-ownership models.

19. DESIGNING POSITIVE PLATFORMS

MARINA GORBIS

In the early days of digital technologies, we did not have user-interaction designers. Alan Cooper, one of the pioneers in the field, wrote an aptly named book lamenting this state of affairs: *The Inmates Are Running the Asylum*. In those days, most of the software interface decisions were made by engineers, and much too often one needed to be an engineer to use their creations. Over time, interaction design emerged as a discipline with a set of rules and conventions, so ordinary people could use many of the previously forbidding tools. We now know where to put the buttons on the screen and how many links to embed so that people can get to the information they need.

Many of today's on-demand work platforms are the beneficiaries of this body of knowledge. They have mastered the discipline of inter-action design and brought it to new heights—when it comes to *consumer* experience. Uber, Munchery, Postmates, and many similar apps are exquisitely designed, sometimes even addictive for users. They make previously laborious processes effortless and seamless. Swipe your phone with a finger and voilà—your ride, your meal, your handy-man magically appear.

But the apps are not only platforms for consumption. They are quickly becoming entry points for work, gateways to people's liveli-hoods. In this sense, whether or not platform creators like it or realize it, they are engaging in another kind of design—socioeconomic design. This involves the design of how people structure their work, earnings, and daily schedules. And here we find ourselves in the same phase as interaction design was decades ago—the inmates are running the asylum. The stakes, however, are much higher; instead of just convenience, we are talking about livelihoods. Herein lies the urgent need to develop on-demand platform design as a discipline, and a discipline that includes not only technological expertise but also the best thinking from disciplines such as economics, political science, governance, and others. Otherwise we risk ceding many key social choices about how we work—what is fair compensation, who owns our work products, data, and reputations—to platform creators. We embed values into our technologies, and today such values are reflections of Silicon Valley's ethos and funding models.

The design of "Positive Platforms"—online platforms that not only maximize profits for their owners but also provide dignified and sustainable livelihoods for those who work on them—is one of the most urgent tasks we are facing today. Cooperative ownership structures give us an opportunity to shape on-demand platforms in a positive direction. After all, the polarization of economic gains between platform owners and those who use their apps to earn livelihoods is one of the biggest dangers in an economy dominated by platforms; distributed ownership can go a long way in remedying this. Platform cooperativism also contains promise of a more democratic governance, with those working on platforms having voice and power to make good economic decisions from the point of view of owners and platform workers. By themselves, however, these levers may not produce the desired outcomes. They need to be combined with careful attention to design elements embedded in platforms themselves.

Platform design choices should arise from the experiences of people interacting with them, including consumers *and* platform workers. To help think about the latter, the Institute for the Future last year engaged in ethnographic research involving people who are working on platforms in different locations across the United States— San Francisco, New York, Miami, Chicago, and elsewhere. We wanted to understand the variety of their perspectives and immerse ourselves in their vocabulary. We recruited study participants with two criteria in mind: the degree of engagement or time spent on platforms (from passively renting to working full-time) and degree of skill required (from Uber drivers to those working on HourlyNerd).

Based on this research, we've begun to identify some principles or rules that should guide designers in order to achieve more positive outcomes for workers:

1. Earnings maximization. It goes without saying that any platform, cooperative or not, should have a viable economic rationale for its existence. In addition, however, platforms should and can be designed to optimize opportunities for those working on them to earn a good living. Connections between design choices and earnings are not yet fully understood. Research has suggested, for instance, that for some types of work people do not do as well financially when the platforms set minimum wages as compared to when workers can set their own wages. Arun Sundararajan and others, as well as our own observations, have found that platforms on which workers can organize their own small enterprises, like Airbnb, rather than those in which workers merely serve the needs of the platform, tend to generate higher levels of incomes for platform workers. Many platforms can go a long way in providing services and feedback loops to help those working on them create more lucrative small businesses.

2. Stability and predictability. We are in a phase of prototyping and experimentation in platform design, a practice that is key to Silicon Valley's style of innovation. But in the case of platforms this innovation has a direct impact on people's livelihoods. Imagine if every month you came to work and your salary were different; this is exactly what many on-demand workers experience today. Participants in our study, for instance, described shifting pay structures with only a few days' notice. Platforms should be structured in ways to minimize such volatility or give workers sufficient time or compensation to adjust to forthcoming changes.

3. Transparency. We need transparency at two levels: at the level of the platform algorithm itself (so that workers understand how to increase their earnings) and at the level of archived data (so that those working on platforms understand how their personal data is being used). Many people we interviewed reported how difficult it was to figure out how to maximize their earnings on platforms due to the general opaqueness of the algorithms powering them. Workers may consequently have trouble calculating their actual hourly wages or whether it is worthwhile for them to take on certain tasks.

4. Portability of products and reputations. Reputation is what powers access to work and ability to earn incomes for those on platforms. People working on platforms should be able to own the products of their work and their reputation histories, and carry them from platform to platform. Platform reputations are often directly tied to earnings as well as opportunities for various types of work. This is how one research participant describes the experience of "losing" a reputation—as well as the accompanying confusion when a platform was acquired by another company: "All of my portfolio links are broken now, and I don't think people can find me anymore."

5. Upskilling. While traditional career ladders may not be relevant in the world of on-demand work, people still look for opportunities to increase their levels of skill and expertise. The best platforms already show those who work on them pathways for learning a particular skill and connect people to resources for advancement. Upwork, for example, not only provides forums for people to mentor and provide support for each other but also links them to free and paid courses where they can acquire desired skills.

6. Social connectedness. Many of today's workers are creating communities outside of the platforms where they work to exchange tips and connect with each other. Reddit, Facebook, Google Groups, and other social media sites are becoming de facto places for this. As one person we interviewed said, "I think it's important for me to build a relationship with the people that I work with." Mechanical Turk workers have come together on a series of forums not only to create a sense of cohesion but also to advocate for their rights. Platform designers can make this easier by enabling and fostering such communities.

7. Bias elimination. Networks are at the core of what makes platforms work. Unfortunately, networks can be exclusionary (due to clustering of people with like minds and backgrounds) and polarizing (because more connected nodes tend to draw even more connections). In formal organizations, decades of labor struggles and court rulings have established some basic rules and principles for non-discriminatory hiring and promotion. We need to evolve such rules and principles in platform environments. Platforms could integrate mechanisms for surfacing bias as well as eliminating it. Models for this come from some recent startups such as Knack, which matches people to job opportunities independent of their degrees or demographic characteristics, or Unitive, which develops software that helps spot unconscious bias in job descriptions.

8. Feedback mechanisms. It is hard to negotiate with algorithms, and most platforms do not have HR departments for handling the issues that those working on them encounter daily, from late payment to unfair reviews. Platforms need to establish feedback mechanisms and equivalents of customer support services for those working on them. "If I were starting an Internet company or designing an app for something," one of our respondents said, "I would say that we must have phone customer service 24/7, and we must be able to guarantee payment." As platforms come to dominate more sectors of the economy, customers and workers alike will come to expect effective means for providing feedback.

These are just some of the early principles we've been able to distill for Positive Platforms. We shouldn't lose sight of the fact that platform design by itself will not ensure sustainable livelihoods. It is just one of the levers, along with governance, ownership, and funding mechanisms. Ownership by itself, for instance, may not guard against natural network biases or unfair reputation systems. This is why, when designing cooperative platforms, we need to think about technology or interaction design along with governance and ownership design. And let's not forget that the platform infrastructure also sits within a larger ecosystem of economic, social, and regulatory frameworks. We need to be thinking about platform design as a part of rethinking this much larger ecosystem as well.

While in the past most value was created and flowed through formal organizations, networked platforms and protocols are becoming the new operating systems for value creation. If before we needed layers of management to coordinate activities—to find the right people, allocate tasks, and share information—increasingly these are being done by algorithms. At least for now, these algorithms are created by human beings, and the stakes for how we design them are high.

20. CONVENIENT SOLIDARITY: DESIGNING FOR PLATFORM COOPERATIVISM

CAMERON TONKINWISE

Designers aim to make things easier, more productive, or more enjoyable. They do this by creating careful interfaces for products, communications, or physical and digital environments. When well designed, in ways appropriate to the contexts in which they will be used, things can:

1. Attract people to undertake certain activities, or to do those activities more frequently,
2. Bring focus to these activities, and
3. Make those activities more habitual.

It is important to see how these three things are related but also how they oppose each other. To attract people to do an activity, the design must fit the activity with people's current habits and expectations. If someone has to figure out new ways of doing something, it might obstruct their willingness to follow through on that activity.

This is important when starting to think about the design of platform co-ops. Sharing has always been an important aspect of being human. But it has been marginalized by transactional, ownership-oriented systems. Today, systems that are nonprofit or cooperative feel less convenient than commercial market offerings. That's why design for platform co-ops needs to be focused on attracting people.

Until now, the design of such systems has leaned toward automating interactions, mostly as a result of the business-as-usual investment structures underlying them. Platform cooperatives resist the extraction of profit that only benefits the few by promoting systems that focus participants on forms of sharing that can enhance the sustainability of everybody in that ecosystem.

How can you design with the aforementioned tensions in mind?

TRADE-OFFS AND OBSTACLES

Some people do things, no matter how effortful, because they believe in them. But many will engage in a new system only if it affords them a chance to get something for little or no money or effort. Currently, successful systems of sharing have taken advantage of smart mobile devices to lower the effort involved: memberships, profiles, and ratings provide levels of assurance about the transaction; payment systems and default communications are semi-automated.

I suspect that sharing interactions should always have at least some of the awkwardness of encounters between peers who do not know each other so well. An added advantage of designing to retain some social effort in the interaction is that it disincentivizes those seeking to scam the system.

In practice, this might mean interfaces that allow a range of payment types. If the on-demand economy already involves paying with contributed labor, social friction, and other costs, then interfaces designed for platform cooperatives should enable even greater variability.

Ride-sharing applications already display different wait times and costs for different qualities of rides. If extended to display other qualities of those offering the rides—unionized drivers, drivers with health care coverage included, drivers otherwise currently unemployed—interfaces could afford convenient forms of solidarity; I choose to wait longer or pay more to benefit a certain kind of driver. Despite obvious privacy concerns, such systems could also be used to discourage other kinds of drivers. But such design systems allow customers to pay more to subsidize particular in-need workers.

MEMBER PROFILES AND PRONOUNS

Sharing platforms promise, on the one hand, ways of exchanging with strangers that are not solely money-based and, on the other hand, they encourage re-establishing exchanges of goods and labor around new forms of sociality, ones that are less confined and more cosmopolitan. In many ways, the core of the political potential of sharing economies lies in the new kinds of social groups that they can sustain. Prior to the rise of digital platform-based labor economies, these groups tended to be formal organizations, like community organizations, cooperatives, or unions. Platform cooperativism presents the possibility that much of the current sharing economy should be restructured around and beyond these existing organizations. But the work required to sustain such structures can limit involvement to those with the time and skills to do so.

The interaction designer for platform co-ops should therefore work to enable participants to create and maintain "lighter," but no less sustainable, communities. At the outset of what is now called the sharing economy, many systems were membership-based: for instance, Zipcar. Current mainstream sharing economy systems require participants to have an online account, but most have dropped the rhetoric of "member." The dominant players continue to brand themselves as a community, while users experience the systems more like customers. There is an opportunity for platform co-op designers to revive the project of establishing genuine community.

A pivotal touchpoint is the ubiquitous profile page. In the socially embedded economies of platform co-ops, these play a crucial role in presenting participants to each other. What information a form gathers about someone, and then how that information is curated by an interaction designer into a profile that others can see, is axial to the nature of interactions people in that platform have, including issues of trust.

In the interaction design of an app, there is often uncertainty about which pronoun to use to describe a user's assets: "my photos," "your music," "Cameron's account." The issue here is not just one of usability, but of how to describe what it means to cooperate through this platform. Platform co-op designers should make more strategic use of "we" and "our," thereby building a sense of collective ownership and mutualism rather than individualism into all aspects of platforms, including the pronouns used to guide participants.

PROTECTING THE COMMONS

Commons are sometimes considered tragic because of the asymmetry between individual and collective cost and benefit. An individual might take the risk of exploiting a common resource because the individual benefit is great, whereas the cost—distributed across many other people—may not be noticed—unless everyone else similarly exploits the resource. Because of this asymmetry, commons need to be negotiated through conventions that are actively maintained.

Cooperatives formalize this need to protect the commons. The interaction designer of more cosmopolitan labor platforms must find ways of encouraging these protective actions without overburdening more distributed participants. The Internet plays an ambiguous role in this respect. On the one hand, the "true identity" and traceability that have resulted from monopolies like Google might ensure that individuals whose actions might be threatening a common resource are identifiable. On the other hand, the anonymity that characterized the first decade of the Internet lowered the costs of criticizing the actions of an exploiter. Today's interaction designers might therefore allow participants in a cooperative platform to be at times identifiable and at other times anonymous or collective.

The on-demand economy relies heavily on peer-rating systems. At the moment these tend to take the form of crude numerical schemes that are known to be ineffective, biased, or simply poorly designed. It is possible for platform co-op interaction designers to make the process of peer-rating a more nuanced interchange. Opportunities exist for the "service design" of interchanges among participants that can protect a commons from degradation.

Designing platform co-ops is an exercise in achieving a new kind of balance—between ease and effort, between individuality and collectivity, and between privacy and transparency. Designers need to get this balance right to ensure a platform that encourages constructive negotiation among members for collective benefit.

21. DESIGNING FOR PRIVACY

SEDA GÜRSES

Taking privacy seriously is part of what it means to create a democratic, accountable, and fair platform cooperative. This chapter explores some of the questions that you, as a platform co-op developer, should be asking as you design your organization and your platform for privacy.

Privacy is not just a technical matter; it is also a social and political one. The tools and architectures you may deploy are not neutral and may pressure the production of the platform in certain ways. These complexities make governance structures that accompany the technical evolution of the platform key to developing a healthy and cooperative privacy design process.

I propose practical ways to organize just such a process. I draw on what I've learned from seasoned platform cooperativists Felix Weth (Fairmondo), Emily Lippold Cheney and Noemi Giszpenc (Data Commons Cooperative), and Alex Rosenblat (Data and Society), who studies how Uber drivers experience their work.

WHO DOES WHAT

In "sharing" economies, the organization of platforms tends to rely on a dichotomy between developers and users—and a presumption that the expertise and decision-making power lie with the developer. For example, TaskRabbit connects "taskers" and "clients," which they refer to collectively as "users"; Uber says they "create opportunities … and improve the way everyone gets from A to B" inspired by the journeys of their "users and drivers." Consistently, developers are depicted as the makers, and users as passive consumers that have to make do with the developers' choices.

Through the developer-user dichotomy, sharing-economy platforms erase the material conditions, the experiences, and the needs of all other parties that contribute to or are affected by the functioning of the platform. They often externalize a number of social, economic, and legal risks to the user, leading to a social sorting of those who cannot bear these risks. For example, these platforms typically establish trust in their workers-disguised-as-users by exposing personal and performance information. This means risks from these information exposures, such as discrimination, are also borne by the users.

You can break out of the user-developer dichotomy by reflecting on your tools, introducing democratic processes, and addressing privacy issues that most platforms refuse to account for.

CO-OP INFRASTRUCTURES AND THE DIVISION OF DESIGN LABOR

The platform, conceived predominantly on Silicon Valley's terms, is not neutral. How, then, do you design a platform for privacy with tools and processes designed for the extractive data economy?

Felix Weth explained to me, for example, "We decided against monetizing user data. That takes away the antagonism and allows us to enter a conversation about how to treat user data." This is commendable but can also be challenging. Is your platform co-op technically equipped to operate analytics and authentication on its own rather than through data-hungry third-parties? Development methods matter, too: will you use agile development practices to create an environment of incessant feature changes? What are some co-op–centric development practices that can enrich the user-centric models?

Rather than reducing these questions to developer dilemmas alone, ask the development team to work closely with the co-op's constituents in rethinking how the platform will be aligned with the values of the cooperative. A co-op may go a step further by explicitly recognizing the ways that different kinds of participants contribute to the advancement of the enterprise. This way your constituents' collaborations and contributions can become integral to establishing the trustworthiness and fairness of the platform, reducing the centrality of individual worker performance and personal attributes as signifiers of trust. These contributions can be formalized in a governance model that includes a privacy supervisory board to promote and assure the accountability of privacy design decisions.

COLLECTIVE INFORMATION PRACTICES AS A PATHWAY TO PRIVACY DESIGN

Once your governance structure is settled, you can iteratively ask these three questions to guide your privacy design throughout the evolution of your platform.

1. **What are the practices and activities that will be mediated through the platform?**

    Identifying the desired collective information practices is central to determining the relevant privacy issues for a platform. For example, should co-op members, clients, and other parties be able to send messages to each other using the platform? If yes, should they be able to broadcast messages to all members? Should these messages be visible to the public, to other co-op members, to clients, or to the platform administrators? Why or why not? Do all parties understand what is visible to whom? The answers to these questions determines which privacy approaches may be appropriate.

2. **What are the potential information flows associated with your collective practices?**

In a naïve design, in order to enable those practices and activities, your platform will collect information about all those who engage with the platform and the surrounding environment. Environmental indicators may include data to evaluate the performance of the technical platform, or metrics that capture the activities of co-op members or clients. Map out and discuss how this information may be collected, used, or shared with third parties. You will refine this mapping based on your privacy design.

3. **Which approaches to privacy design are appropriate for the different practices?**

This is the big one. Within it, consider the three approaches that follow.

The *privacy-as-confidentiality* approach offers publicly vetted techniques that will allow platforms to minimize data collection and avoid single points of failure. Offering members end-to-end encrypted private messaging, for example, provides the co-op a protection against coercion or unreasonable search-and-seizure requests from law enforcement. Techniques based on encryption and secure protocols may be indispensable for guaranteeing that platform administrators cannot single-handedly compromise sensitive information related to practices such as voting, anonymous participation, and reputation systems. One approach is to use trustworthy third-parties to run polls or elections, in which case you should take measures to protect the data collected by the third party and ensure the reliability of the results.

Privacy-as-control approaches focus on access control and transparency. These include measures that the platform can implement to assure accountability and compliance with data-protection requirements. Such mechanisms can also be used to provide platform users with choices about data collection and processing.

Co-ops may face tensions due to conflicts among legal requirements. Data protection laws require that platforms make transparent what data they collect, process, and share, and provide controls over those flows. In contrast, co-op laws may require that members have access to a detailed member roster. If you publish the roster, you may inadvertently subject members to spamming, harassment, and doxing, and risk the abuse of your co-op by commercial competitors. These risks will most forcefully impact those who do not possess the social, technical, and legal capital to bear the costs of those risks. Further concerns may be raised when a third-party service is used, like Google Analytics or Facebook Groups. Discuss with your co-op different design solutions that allow constituents to control which information is disclosed to whom while remaining compliant with transparency requirements.

If your platform is also used to facilitate work, then the platform potentially needs to comply with laws about workplace surveillance. Privacy settings and the ability to log off without continued tracking, as well as mechanisms to examine and dispute the data the platform tracks, may be vital to providing workers a fair workplace.

Privacy-as-practice approaches involve design principles aimed at respectful and accountable interactions among all co-op constituents. Subtle decisions regarding whether legal names will be required, or if anonymous participation is possible, may impact who is able to speak and how they are held accountable for their actions. Design principles like "social translucence" can provide creative alternatives for linking workers and clients without exposing their personal attributes. A code of conduct is also essential and should accompany technical mechanisms—describing platform norms, promoting a safe environment, and outlining procedures for due process.

It is crucial to privacy-as-practice approaches to design the platform in such a way that makes information flows, and their potential consequences, intuitive. If fulfilling a certain task has an effect on a worker's reputation, or if an algorithm filters work bids based on select criteria, this causality should be obvious to the parties concerned.

22. HOW CROWDFUNDING BECOMES STEWARDSHIP

DANNY SPITZBERG

Crowdfunding can seem ideal for building cooperative platforms on the Internet.

Intuitively, this makes sense. Desperation and necessity inspire many of us to form co-ops. And because co-ops can only accept non-extractive investment, crowdfunding can look like a great way to start—a digital barn-raiser that builds community without tapping it out. In practice, however, many co-ops struggle with crowdfunding. I believe this is because marketing has skewed our view of crowd-funding by influencing how we think and feel about community.

What does "community" really mean here? Community is collective action with a shared story. We join clubs, co-ops, and campaigns that offer material benefits—things that matter to us on a daily basis—and we stay because of solidarity with our peers and a purpose we can achieve together. The more we act collectively, the more we strengthen these incentives.

Incorporating as a co-op is a long way from building community. While there is a grain of truth to the idea that co-ops are the orig-inal crowdfunding, people experience co-ops through organizing and campaigns, not bylaws or business plans. More important, we can't extract generosity. That is what marketing tries to do in platform cap-italism. However, we can form relationships rooted in reciprocity and generosity through cooperative arrangements.

I learned these lessons last year, when I partnered with Loconomics to crowdfund their platform and grow their membership. On paper, Loconomics had a beautiful model: a local services co-op owned by the freelancers doing the work. The user-owners get tools for booking clients, a growing marketplace, and a dividend based on the co-op's performance. But the appeal of joining a co-op needed as much validation as the platform itself.

To research needs, I interviewed a representative group of a dozen freelancers—some with their own client base, and others finding odd jobs on platforms like TaskRabbit. Nobody felt misinformed, much less exploited, with what they get through on-demand service platforms. However, they craved the feeling of belonging to something bigger. A part-time plumber with a philosophy degree described the ideal as "less a client base, more a partner base"—in other words, a co-op. But would anyone pay to join one?

People give endless feedback on ideas, but only commit if they see value. A sure way to make this shift is through opportunities for people to test a prototype and express their emotions.

EMOTION WITHOUT EXTRACTION

"All emotion is involuntary when genuine," according to Mark Twain's casual wisdom. I believe this rings true for anyone building co-ops, maybe more than the first cooperative principle of "voluntary and open membership." And for anyone who has run a crowdfunding campaign, mobilizing genuine emotion can sound like difficult, draining work.

After working in dozens of campaigns, I've seen a tension play out between crowdfunding and membership. Crowdfunding is a one-off moment of collective action, but when the projects that we care for also take care of us, people come together and stay together.

How might we reinvent crowdfunding so that collective action continues?

It's tempting to search for answers on the Internet. But before going online, consider the case of a real-life forest. Neera M. Singh, author of a 2014 forest conservation study in Odisha, India, found a region that challenges the logic of paying individuals to manage resources as market goods. She observed how villagers harvest only what wood they need from the forest, and sing songs celebrating its cool breeze, too. Singh concluded that community stewardship sustains thousands of villages because people organize their labor both *effectively*—forming accountable relationships around their work—and *affectively*—developing shared identity in the process.

The story of stewardship in Odisha shows another side of crowdfunding. While starting a project might depend on pooling financial contributions, sustaining it requires emotional investment.

Query your favorite search engine for images of "women laughing alone with salad," and you'll see a cliché used to evoke health and happiness. I suggest taking a look if you haven't recently—partly because it's hard not to laugh at the fake emotions, but mainly because a similar caricature shows up in how on-demand service platforms market themselves.

TaskRabbit, for example, portrays images of smiling helpers cleaning kitchens while women hold babies. Unlike stock photos, however, we meet TaskRabbit in real life. Their marketing may be full of clichés, but on-demand service platforms are also full of opportunities for us to become emotionally invested.

Platforms like TaskRabbit leverage our emotional investment to grow their user base. Their user experience is designed to delight us, especially at key moments around transactions. When interacting with a chef, host, or any service provider who loves their job or gig, we enjoy acts of kindness that have little to do with rating systems. But platforms do not support self-organizing. Instead, they leverage community activity to increase user engagement, and resist attempts to leave. TaskRabbit charges $500 if you move any consumer-provider relationship off its platform.

This is the norm in platform capitalism: products extract value from transactions for outside investors. The platforms connect us to resources more than they operate as a resource themselves or a place to gather. In this context, our emotions are more like "laughing alone with salad" and less like singing together in Odisha's forests.

The real issue with emotions lies with the conditions in which they are extracted. Emotions on the Internet can be better understood with Arlie Hochschild's theory of "emotional labor," which describes how we adapt our emotional expressions in deep and superficial ways to align with workplace rules. While Singh found villagers laboring happily, defying market logic, Hochschild argued more than thirty years ago that emotions get commodified in a capitalist service economy.

Looking at how emotions change over time, however, shows how people become invested. Elizabeth Hoffman's 2016 study of worker co-ops found that embracing emotion ultimately benefits democratic participation. As individuals get comfortable expressing themselves, they develop an identity as co-owners—their workplace and co-workers feel like "home" and "family."

Such transformative, humanizing experiences contrast with how we relate to one another through marketing. These are also how investment grows into stewardship.

BARN-RAISERS FOR STEWARDSHIP

Happily, my partnership with the Loconomics team ended with their focusing on community before launching a product. To see what invitation attracted people most, they swapped their full website for a simple sign-up page. And to learn about user experience, they welcomed service providers and clients to events where they could try the app, volunteer, or become owners. Getting together finally made it possible to experience what a community might feel like.

At a minimum, community is a shared feeling of belonging. These feelings well up when people come together, through book clubs and parties, and they evaporate when the organization shuts down, puts up a pay-wall, or simply has a change of heart. This precariousness is easily overlooked, however, when a platform manages to balance user satisfaction and extraction.

Building community through crowdfunding plays out in a similar way. It starts with a goal of mobilizing contributions from many individuals. With enough incentives and excitement, the possibility of passing a funding threshold triggers collective action. This usually happens only once. Very few campaigns lead to what Hochschild calls "deep acting"—our genuine emotions at work. Most campaigns fall back on "surface acting," the kind of behavior associated with fake smiles. These campaigns strain volunteers, scare supporters, and fail at their goals. And if a project *does* get funded, any future collective action depends on whoever owns and controls the value created. Without emotional investment in a cooperative arrangement, campaigns run the risk of ruining relationships over unmet expectations.

For crowdfunding to become stewardship, we need rolling barn-raisers—regular activities in which guests can co-create with the gifts they bring, celebrate their accomplishments, and build again.

Marketing strategies extract generosity by developing an audience, message, and call-to-action, leveraging one-way relationships. A barn-raiser is an organizing strategy for a cooperative alternative that involves people, invitation, and engagement (think p-i-e):

- Connect with people. Audiences are passive, but people put emotion at the core of cooperation. Learn who might join the effort, and what they're trying to get done.
- Make an invitation. Messages are static, but invitations cultivate voluntary and open membership. Define what you want to celebrate, together—in person or online.
- Sustain engagement. A call-to-action limits inputs, but engagement supports democratic ownership and control. Seek participation more than financial contributions.

By starting small and learning along the way, barn-raisers can "grow the pie" for co-ops. This is how crowdfunding becomes stewardship: raising expectations, embracing the challenge, and sharing the value as community grows.

23. ECONOMIC BARRIERS AND ENABLERS OF DISTRIBUTED OWNERSHIP

ARUN SUNDARARAJAN

In May 2015, I chatted with OuiShare's co-founders Antonin Léonard and Benjamin Tincq during their OuiShare Fest, the annual gathering of over a thousand sharing-economy enthusiasts in Paris. I sensed a tension at the Fest between the purpose-driven and profit-driven participants: those who saw the sharing economy as a path to a more equitable and environmentally sensible world, and those who were excited by the massive infusions of venture capital into hundreds of burgeoning sharing-economy platforms.

Léonard spoke of the confusion and disappointment he detected from those who had hoped that the sharing economy would really change the world. "And because there was so much hope, the ones that were once so hopeful are now so disappointed, in a way," he said. "But maybe the problem is not so much how much money was invested, but why did we have this hope?"

Tincq, while agreeing with the perception of growing disenchantment, was focused on a simpler point: that the shift away from purpose and toward profit was driven primarily not by a change in philosophy but by a need for growth capital. In his view, at the time, for a nascent platform to bridge the early-stage gap and get to critical mass, there was no practical alternative to venture capital.

Tincq's point echoed a theme from a panel discussion I had organized about new ownership models at the 2014 Social Capital Markets conference with Janelle Orsi, Lisa Gansky, and Adam Werbach. A frequently raised theme was how the model of corporate ownership lends itself naturally to the acquisition of large amounts of capital in exchange for outside ownership. In particular, Werbach, a long-time social entrepreneur and co-founder of Yerdle (a shareholder-owned platform that facilitates the ecologically responsible exchange of household assets using a virtual currency), reflected on the challenges he faced looking for ways in which he could structure Yerdle as a cooperative while still preserving the ability to raise the external financing he knew would be necessary to realize his vision.

In light of these and other conversations, I find the excitement about platform cooperatives—especially in the form of sharing-economy platforms owned by their providers and funded through mechanisms other than institutional venture capital—both inspiring and contagious. Presented with these new possibilities, it seems instructive to examine why worker cooperatives mediate a relatively tiny fraction of economic activity in the United States today. There were over 30,000 cooperatives operating in 73,000 U.S. locations in 2009, holding assets over $2 trillion, and generating revenues of over $650 billion. While this scale is not trivial, it is dwarfed by the corresponding success of shareholder corporations. The Fortune 500—the five hundred largest corporations in the United States—collectively generated over $12.5 trillion in revenue in 2015, and the total 2009 cooperative take of $650 billion is less than the corresponding sum of the revenues of just the two largest corporations, Walmart and ExxonMobil.

Economic theory suggests that worker cooperatives are more efficient than shareholder corporations when (1) there isn't a great deal of diversity in the levels of contribution across workers; (2) when the level of external competition is low; and (3) when there isn't the need for frequent investments in response to technological change. This is one reason why a U.S.–based worker cooperative like Sunkist (formerly the California Fruit Exchange, an entity that has, since 1893, been entirely owned by citrus fruit growers), has thrived, where other types of cooperatives have failed to emerge at scale. Perhaps the businesses that have fueled much of the world's economic growth in recent decades have instead been in highly competitive industries, leveraging specialized high-variance talent and requiring large technological investments.

But if one thinks about it, today's sharing-economy platforms do exhibit some characteristics in common with Sunkist, and a worker-owned equivalent to Lyft and Uber seems quite feasible. Point-to-point urban transportation is a fairly uniform service in an industry with a limited amount of competition. Once the technology associated with "e-hail" and logistics is commoditized, which it will be, the economic fundamentals for the emergence of a platform cooperative would appear to be in place.

More important, the network effects associated with ridesharing are geographically concentrated. Thus, unlike platforms such as eBay and Facebook, the barriers to entry posed by an incumbent platform may not be onerous. True, passengers gravitate toward the platforms with more drivers, and vice versa. However, these effects are localized. Most potential passengers in New York care little about the scale of a platform in Los Angeles or Minneapolis. They want the service that has the densest supply in their own city. Furthermore, it is relatively simple for a driver to "multihome," or be a provider on multiple platforms. In other words, each local market is contestable. The same is true for many labor platforms, including those that provide domestic work and home services.

As a consequence, instigating the emergence of a platform cooperative doesn't involve getting millions or billions of users to switch simultaneously. Rather, it might be seeded simply by signing up a few thousand providers. One such effort under way as of the writing of this essay is Swift in New York, a nascent ridesharing effort that hopes to organize as a driver cooperative.

Despite these relatively low barriers to entry, a collective that hopes to build a scalable platform business with a cooperative ownership model faces other challenges. During a panel that Juliet Schor and I participated in at the Platform Cooperativism conference, Schor highlighted an issue her research had uncovered about sharing economy cooperatives: that their value system was often better articulated than their value proposition. Put differently, cooperatives tended to focus too much on how the value would be shared rather than on a compelling offer to create the value in the first place.

Perhaps part of the solution will come from the possibility, created by blockchain technologies, of "distributed collaborative organizations," or DCOs—new decentralized collectives that, in the eyes of pioneers like Matan Field of Backfeed and Vitalik Buterin of Ethereum, can use rules embedded in computer code to align the incentives of different contributors, of financial capital, of expertise, of labor, and of participation. These DCOs are connected intellectually to a variety of related decentralized ownership models. They range from the FairShare Model of Karl Sjogren, which proposes a structure of different classes of ownership shares for different contributors—for founders, people with a continuous working role, for users, and for investors—to the Swarm approach to "crypto-equity" crowdfunding developed by Joel Dietz. If the rules for equitable value distribution are well defined, generally accepted, and become "normal" in the same way that employment for salary at a shareholder corporation was in the twentieth century, perhaps the providers can then focus more of their efforts on creating value.

However, as groups of motivated providers address the challenge raised by Schor, and as experimentation and the quest for normalcy across different platform cooperative and DCO models continues, redistributing platform value using a more familiar route—stock ownership—seem like a promising near-term prospect. In the United States, employee stock ownership programs, or ESOPs, that share ownership with employees by allocating stock to them are quite common. ESOPs create joint ownership as well as a form of profit sharing. And the scale of an ESOP can be quite significant. For example, in 1995, the United Airlines ESOP owned 55 percent of the company.

Creating similar "provider" stock ownership programs—under which providers are allocated shares in a platform—seems quite natural. An early example of a platform that aims to do this is Juno, a ridesharing service started by Talmon Marco, the independently wealthy founder of the messaging company Viber. Juno has committed to ensuring that its drivers own 50 percent of the company's founding stock by 2026. And the scale of such wealth division need not be as absolute. In early 2016, Managed by Q, a labor platform for office services, allocated 5 percent of its equity for its providers.

The parallel with United Airlines seems especially relevant because its ESOP emerged as part of a negotiation between organized labor and management, at a time when the company's survival was threatened by pilot unions. Sharing-economy platforms rely heavily on motivated providers to maintain their brand by delivering a consistent high-quality service experience. Granted, the prospect of automation will weaken the clout of labor in the long run. But in the near term, different digitally enabled provider collectives, manifestations of the idea of "new power" from Jeremy Heimans and Henry Timms, will likely increase providers' bargaining power. As this happens, more widespread provider stock ownership programs may well be a natural response, and perhaps the most pragmatic prospect for sharing the wealth of the sharing economy.

24. THERE IS PLATFORM-POWER IN A UNION

RA CRISCITIELLO

Venturing into the new world of platform cooperativism, to those of us in organized labor, has felt a lot like making a deal with the devil. Our new partners in the tech industry undoubtedly feel similarly about us.

It is no secret that unions are dying. Private sector unions have dismal density, and public sector unions are not faring much better. Attacks in many forms—including what we've seen with court cases like *Harris v. Quinn*, and now in *Friedrichs*—are legal blows that highlight the forces assembled against workers. Organized labor has largely retained its familiar tactics and worldview despite the reality that the economic structures of employment have been turned on their heads. If unions cannot solve labor's woes, it may not be simply because organized labor is dying, but rather because organized labor needs to change.

That is why members of SEIU-United Healthcare Workers West, in partnership with a tech startup, are collaborating to create a worker-owned cooperative of licensed vocational nurses, or LVNs, who can be dispatched on demand to patients' homes through a mobile or online device. Despite being skilled workers, LVNs in California—a workforce largely of women and immigrants—have seen their work prospects diminish, partly as a result of registered nurses expanding their own scope of practice. LVNs can give vaccinations, treat wounds, and deal with most low-acuity issues. But they need new tools to find clients and secure livelihoods. The tech company they're working with is a separate team of entrepreneurs who see the need, as the worker-owners do, to create an on-demand technology that values workers as much as it values investors or the technology itself.

There are two paths emerging at the intersection of organized labor and cooperative employment models. Both address the realities of an increasingly casual and insecure economy that offers workers scarce or nonexistent benefits.

One is an industrial model of cooperative organizing. It looks at how workers' rights and protections are diminishing—benefits like workers' compensation, health insurance, training and development, retirement savings, and sick leave—and tries to fight that tide on scale. This means creating a structure where those benefits are not, as they historically have been, contingent on a worker's employment relationship with a particular employer. The Freelancers Union, for instance, aggregates the self-employed worker community into a visible industry. Aside from the sheer naming and negotiating power that comes with uniting formerly disaggregated workers, affiliating with an organization enables a self-employed worker to gain access to things like group-rate benefits.

The second path is smaller-scale—fashioned after pre-industrial guilds. Nursing within the LVNs' scope of practice lends itself to the guild model, where formal licensure and training is needed, and years of apprenticeship and job placement can help advance the profession. The worker cooperative is building the type of labor market that its members want to see. As the members say in their founding mission statement, their goals are "high-quality, convenient health care on-demand, to grow the LVN profession as well as employment opportunities for highly-trained LVNs, and to increase access to care."

By monopolizing the labor supply in a particular narrow market, organized labor can use the union worker cooperative model to enable workers to own their own labor and enjoy portable benefits, thanks to a collective bargaining agreement between the cooperative and the union. The cooperative guild can start on a manageable level by restricting the co-op work to one classification (like the LVN), but it can later scale to include multiple job classifications.

What both of these paths signify is the potential for value when organized labor and worker cooperatives team up in the "gig economy." Together, they can protect workers' rights while also embracing the flexibility both in workers' lives and in consumer demand that increasingly seems to be the way of the future. On the ground, this has meant listening deeply to workers and being honest about how employment will continue to change.

During an early board meeting, one LVN described feeling scared that her phlebotomist coworkers would soon be out of a job if health care careers that require only a few months of training become "on-demand-ed." The truth is that she may be right. Healthcare workers who do skilled work but whose positions require less than a year or two of training will likely see their work leave the confines of hospital walls. The LVNs are developing a platform worker cooperative to get ahead of precisely that trend. They see change coming, and they want to be in the driver's seat so they can make sure that workers' rights are protected.

The LVN worker cooperative is not just about sharing ownership, governance, and profit. It's about worker-owners controlling their own labor.

The need for startup capital has become complicated, however, given the rightfully uncompromising standards of platform cooperativism. In order that early investors and others oriented toward profit maximization don't gain undue control over the LVN co-op, the members and their startup partner may need to negotiate an agreement; they need to set up guidelines for the relationship between them, for instance, and for how they contract with outside entities. In the process, the co-op needs to reconcile its own democratic structure with the venture capital model that is financing the startup. Neither the co-op nor the startup can succeed without the other, so any success or profit should be shared by both.

When members of the co-op and startup met with a venture capital firm early on, the LVNs described feeling how the investor's drive to extract capital could easily become antithetical to the co-op model, which should allow members to decide how best to reinvest their earnings. Several potential venture capital partners explained that they would require a significant role in the co-op's governance as an assurance for their investment. The co-op board has struggled with what amount of outside control it should be unwilling to give up. Democratic, one-worker-one-vote principles feel, at a gut level, at odds with the capital that the platform needs to grow. Through its bylaws, for now, the co-op has carefully drawn lines around democratic decision-making and reinvestment. It may go without saying that union members are used to threats from those who want to eradicate or exploit them, so they know how to guard the castle.

What unions offer platform cooperatives is possibly the greatest remaining power of any union: the ability to leverage collective power. This may mean the collective buying power to purchase portable employee health care insurance on scale, or the capacity to grow a skilled workforce by providing training and apprenticeship opportunities, or the possibility of creating a worker-centered marketplace like the LVNs' platform. If worker cooperatives and platform cooperatives are the employers of the future, also, union revenues will come from a different source: the cooperative worker herself. The new model moves unionized labor away from entrenched us-versus-them labor relations and lets workers take power directly instead of negotiating for it.

One thing is very clear: all the stakeholders in this new model are taking a leap of faith, trusting that our shared vision of a new employment model will come close enough to satisfying the ethical imperatives of each group. So here we go. There's a little bit of devil in all of us, but maybe we can still cooperate.

25. MAKING APPS FOR LOW-WAGE WORKERS AND THEIR NEIGHBORHOODS

SASKIA SASSEN

How can digital cooperative platforms contribute to better work lives of low-income workers by addressing the specific needs of these workers, at their workspace and in their neighborhoods? What I describe here is inspired by the concept of platform cooperativism: an ecosystem of apps that could address common needs of low-income workers and their neighborhoods.

Such an ecosystem of apps would be one more step toward mobilizing localities around initiatives concerning both workplace and neighborhood issues. This matters, given settings where hardships and losses do not always facilitate trust among neighbors. One way of thinking about my argument is that it is a search for material conditions (e.g., access to cooperative platforms within and across neighborhoods) upon which more complex non-material relationships can be built (e.g., backup support at the workplace, and more aspirational aspects such as trust and solidarity).

The high-end worker is already a full and effective user of these technologies; in the United States, most digital applications have been geared toward high-end workers and households, and to scientific collaboration. Very little has been developed to meet the needs of low-income workers, their families, and their neighborhoods. This is a bad and sad state of affairs given the needs of these workers and families, especially since the data indicate that they have digital access and are willing to spend on apps; but they are mostly confined to mass-market goods, notably music. We also know that digital access is overwhelmingly through their phones. We need more innovations that meet the needs and constraints of low-wage workers. Platform co-ops like Coopify are stellar examples of that.

TRANSFORMING THE NEIGHBORHOOD INTO A SOCIAL BACK-UP SYSTEM

My argument and proposal regarding low-wage workers is the extension of digitization to the larger space within which these workers operate: not only the workplace narrowly understood, but also, and very important, their neighborhood. Apps for low-income workers and their neighborhoods can become part of the larger ecosystem of platform cooperativism. This is already a fact among high-end workers: digitization has become a way of restructuring the connection between work and home. It is inconceivable today that the high-end worker can or does simply leave it all behind when closing the door of her office for the day—on those few days every week when s/he might actually work in the office. We might say the correlation for the low-wage worker is that it is a fiction that s/he can simply leave it all behind when s/he closes the door of her home and goes to work.

The home and the neighborhood have long been support spaces for the working class. Today this is rarer, mostly due to changes in the condition of low-wage workers. Digitization can help rebuild some strength in these spaces. For instance, in the case of trouble (a sick child of a parent who is at work, police violence, etc.) an app on all residents' phones can enable quick deployment of stationary neighbors— grandmothers, hairdressers, and shop-keepers. This is also a first step toward greater neighborhood integration and expanded use of diverse digital capabilities.

UNDERUTILIZATION OF DIGITAL TOOLS AND APPS IN LOW-INCOME NEIGHBORHOODS

This underutilization is a sharp contrast with the case of high-end workers. It constructs a radical differentiation between workspace and neighborhood for low-wage workers. This is disabling and adds to the difficulties in their daily life at and off work.

We must ask what can we do with current technologies but are not doing because of diverse reasons: lack of resources, lack of motivation, lack of interest in low-income households, individuals, and localities, and so on. Important, and too often overlooked, is that the types of applications that are being developed mostly do not address the needs and limited resources of low-income workers, their households, and their neighborhoods. That is why platform co-ops should rethink their value proposition.

In a recent overview, the Pew Center found that 45 percent of U.S. households with less than $30K per year and 39 percent of those with $30K–$50K use mobile phones as their primary digital access. Email at home is rare, and often relies on low-bandwidth dial-up. Researching the use of digital technologies by women across the world on behalf of the United Nations Development Program, I found extensive use of mobile telephones by modest-income and poor women in Africa: it allowed them to run their businesses, mostly diverse small-scale trading.

USEFUL APPS FOR LOW-INCOME WORKERS AND NEIGHBORHOODS

Several efforts are beginning to address these needs. Here are a few examples of mostly recent applications geared to modest-to-low-income households and neighborhoods. Kinvolved is an app for teachers and after school program leaders that makes it easy for them to connect to parents in case of a student's lateness or absenteeism. Many poor neighborhood schools lack easy communication with a student's home; this has allowed self-destructive conduct to worsen, damaging a student's chances for a job or acceptance to college. This app is simple and straightforward: when a teacher, or a coach, or whoever is part of the student's adult network at school, takes attendance or sees something of concern, the family is immediately notified via text messages or email updates—whichever they prefer. The low-income worker knows that if there is trouble s/he will be alerted.

Another app, developed by Propel, simplifies applying for government services, a notoriously time-consuming process. Now there is the option of a simple mobile enrollment application. Yet another such application is Neat Streak, which lets home cleaners communicate with clients in a quick non-obtrusive way. There is also a money-management app for mobiles that combines cash and loan requests, again simplifying the lives of very low-income people who need to cash their paychecks before payday, and can avoid the high interest rates charged by so called "payday sharks."

A very different type of app from the aforementioned is Panoply (presented by Robert Morris): an online intervention that replaces a health professional with a crowd-sourced response to individuals with anxiety and depression. What I find significant here is that it has the added effect of mobilizing a network of people, which may be one step in a larger trajectory of support that can also become a local neighborhood network. Panoply coordinates support from crowd workers and unpaid volunteers. It incorporates recent advances in crowdsourcing and human computation, enabling timely feedback and quality vetting. Crowds are recruited to help users think more flexibly and objectively about stressful events.

Another useful tool seeks to develop new ways of working together online. This is something quite common among middle-class users and in certain professional jobs, but far less likely among low-income workers. But it could be useful to the latter; it can enable a sense of the individual's worth to a network ("I matter to my community"), and thereby feed solidarity and mobilization around issues of concern to low-income neighborhoods, families, and workers.

NEW CHALLENGES THAT CALL FOR NEIGHBORHOOD COLLECTIVE ACTION

Neighborhoods are important spaces for low-wage workers. In the past they often enabled union organizing and the formation of mutual-assistance organizations. Much of this is lost today. There is work to be done to strengthen this neighborhood function. But this can only happen if the neighborhood is a space for connecting, collaborating, and mutually recognizing each other. Given the development of apps geared to low-wage workers, platform cooperativism could enable significant scale-ups in the deployment of such apps and in their spread. One key mode of scale-up would be shared ownership and shared governance. This would have the added effect of enabling collaboration among workers and among residents within and across neighborhoods, joining hundreds of years of the history of cooperatives with the digital economy. I see here beginnings of possibly new social histories.

26. THE CROWD: NATURALLY COOPERATIVE, UNNATURALLY SILENCED

KRISTY MILLAND

When you think of a crowd, you'll likely envision large numbers of people, many voices, and the wisdom of a variety of viewpoints. In crowd work, all of these features are what makes the platforms so useful; work is done quickly by many with a variety of skill sets and abilities, and their group wisdom leads to a high-quality product. It only makes sense that the future should bring us self-governing crowds, those who run their own platforms to ensure that their needs are respected equally with those of the businesses that leverage their labor for a profit. To date, however, not only has no such collaborative crowd work platform emerged, but the voice of the workers themselves has mostly been ignored in the discussions about creating these platforms. If we are to move forward into a future of labor where many existing jobs are displaced by robots or algorithms, relegating us to work on crowd platforms, we must design online workplaces that are run collaboratively, or we will all be beholden to exploitative companies who do not have the workers' best interests at heart.

As a worker on Amazon Mechanical Turk (mTurk), I know insult and exploitation firsthand. For example, this week I decided to set aside an entire day in order to make as much money as possible. Out of the eight hours I spent on mTurk, I was able to complete 166 tasks, called "HITs," and I earned only $19.64.

The first problem is how much effort it takes to find work; I can't just restrict myself to *good* work, I have to instead do *any* work that is available in order to make money at all—but even that is limited as I'm Canadian and most HITs are qualified only for U.S. workers. This discrimination makes no sense for most work on the platform, since Canadians can tag photos, categorize sentiment, or transcribe text just as well as their American neighbors, but I'm excluded regardless.

Next, accepting just any HIT exposed me to some horrific content, such as a survey with at least eight videos of either really happy or really sad or violent content, the title of which was something along the lines of, "Does this make you cry?" It took me at least an hour to recover emotionally and physically from that HIT alone, which means that I had an entire hour of wasted time. The pay? $2.75 for 50 minutes. Had I realized what the content was about and how long it would take, I would have never accepted the HIT, but the content was not fully described upfront. In the past, I have been faced with HITs that included ISIS murder videos or animal abuse, or worse. This is what I have to do to generate income as an mTurk worker, and it is damaging to my soul. Yet if I have no other avenue for income, it is my only option.

One of the platform's most egregious abuses is allowing workers to go unpaid for work completed to the best of their ability. Amazon not only condones this wage theft but has made it a feature, since the employer who posts work gets to see what is submitted in order to adjudicate it. All employers have to do is reject the worker, denying them payment, and they get to keep both the work and their cash. It is scraping the bottom of the barrel when a worker not only has to face being paid pennies per hour for their hard work, but also the possibility of not being paid at all.

Many people assume that workers such as myself are all from developing countries, are unskilled, barely speak English, have no education, and will cheat to steal money from those who post work on the platform. It gets worse, with comments about the fact that we are literally the unwashed masses in our pajamas doing work for pennies an hour, the lumpenproletariat so clueless that it is a favor to pay us even a pittance in order to give us any job at all. This is utter garbage, as studies clearly show we are highly skilled and educated workers (see "Amazon Mechanical Turk (mTurk) workers are highly educated" at TurkerNation.com), the bulk of whom come from the United States. We are now awakening to a class consciousness. We are beginning to push back against the drive to put us to work so unethically.

In attempting to speak out about the horrors we face on the job, we've realized that those outside the system have no idea what is going on inside. Journalists seem oblivious to the fact that while a robot may not yet be able to perfectly replicate the writing style of a talented human author, a group of humans can, and cheaply. The job of writing for the news is likely to be one of the first to fall to crowd work, but the field will quickly eat up other careers as well. Crowd platforms are being tested to replace doctors (CrowdMed), software developers (TopCoder, InnoCentive), graphic designers (99designs), interior designers (CoContest), and more. There are few jobs that require a degree of skill or education that could not be matched by the crowd.

If we allow the status quo to march ahead, with platforms that are designed to support abuse and no legislation to stop worker exploitation, then it won't just be the lumpenproletariat screaming for change. It will be you, your mother, your sister, your daughter, and everyone you know and love being paid pennies an hour—if they are lucky enough not to have their work rejected or stolen. As Torben Schenk, a critical political economist and an advisor to a member of European Parliament, put it at a recent workshop, we must view the current field of crowd work as a juggernaut plowing downhill, and we can no longer jump out of its way.

I don't claim to know what the perfect platform design is, but when designing for a crowd, one must engage the crowd in active dialogue in order to find the answer. At the Platform Cooperativism conference, the intent was for a widely varied group of people to talk, generate knowledge, and move forward to take action. Many loud voices were heard, and others lurked in the background taking notes. But one sad fact I faced firsthand was that the crowd workers were not there in force. On top of that, at the Worker Voice panel, where Karla

Morales and Zenayda Bonilla discussed how their own cooperatives operated, and I described my experience with mTurk, there were very few people in the audience. I attended many other panels and talks, and it seemed like the audience was hungry to hear how they could create their own cooperatives, but they mostly looked to me like experts in business or law, or CEOs of cooperatives, although that seems like an oxymoron to me. This must end here.

While you can copy a blueprint already created, and must consider legalities as you strive to form, there is no voice more important than that of the workers. Without workers you cannot have a cooperative. I would hate to see so much effort and time put into developing a platform, only to have it fail because it does not fit the needs of the workers it was expected to serve. If you want to know what will work, ask those who will work on the platform.

It is clear that the labor platforms we have now aren't working for those who use them. Features such as easy communication and protection from exploitation are ignored in favor of isolation and wage theft, while the companies that run the platforms continue to turn a profit. Even though it may only be a few million people who are working full-time on such platforms right now, the speed, low cost, and ease of getting work done through the crowd means careers will disappear one by one as the crowd takes over those jobs. No matter who you are, you will see at least parts of your job taken off your desk, and that means portions of your income will disappear along with them. We must stand together and create places of work that allow the crowd to set its own standards, enact its own protections, and alter the future of work to be balanced between worker and employer rights.

We cannot do so without listening to the voices of those who are already using such platforms—both the crowd and those who leverage them. If you want to protect the future of labor by protecting the laborers, the time to put workers' expertise at the helm is now. Dystopian or not, a future in which we are all forced to work on something like Amazon Mechanical Turk is not out of the picture unless we change the frame—by creating the competition ourselves.

27. PLATFORMS AND TRUST: BEYOND REPUTATION SYSTEMS

TOM SLEE

Do new technologies embed a set of values?

Some contributors to this book argue that platform cooperatives can clone and change the ownership structure of existing platforms. This assumes that the technology stack is essentially neutral; it can be operated for private profit or it can be operated for cooperative goals, at the behest of its owners. For others, the technology stack is not neutral. Embedded within it are the values of the project. Even if technology does not determine outcomes, technologies have "affordances" that favor some outcomes over others. According to this view, the networked structure of the Internet has certain affordances for democratic politics and decentralized organization; the use of free and open source software is a political commitment to openness as well as a technological one. The enthusiasm for the blockchain—the distributed, decentralized ledger that provides the basis for Bitcoin and other digital currencies—is a recent expression of the same idea, even if the particular values are different.

Consider, for instance, rating systems. Many see rating systems as another technology that embeds a democratic and egalitarian politics. Rating systems have become an alternative to experts. No longer do we have to rely on an elite class of old-guard establishment critics to guide our tastes: we can do it ourselves. We rate books on Amazon, films on Netflix, restaurants on Yelp; it's the democratization of criticism and recommendation.

Reputation systems are a special case of rating systems, in which people rate other people, conferring on them a reputation. Some see reputation systems as the primary innovation of the sharing economy. We can get into strangers' cars, eat at their tables, stay in their homes, or lend money to people we will never meet—all because ratings-driven reputation systems seem to have solved the problem of trust between strangers. For their proponents, these systems come with particular embedded values of democratization and decentralization: Tim O'Reilly, for instance, believes we may be entering a new era of "algorithmic regulation," in which reputation systems replace credentials and inspections once provided by public agencies. We don't need cumbersome regulations to ensure good behavior, we can do it ourselves.

The promise of reputation systems, and the claim that they democratize trust, are largely mirages; they do not embed values that some think they do.

Even though reputation systems look and feel like product-rating systems, it turns out that we act differently when we rate people to when we rate products. The difference shows up in the rating distribution: on Netflix or on Yelp, reviews on a five-star scale show a peak at a value between three and four stars, and tail off to either side, with a smaller but significant number of ones and fives. On sharing economy platforms, most ratings are in the four-star or five-star range. We can call these "Lake Wobegon systems," after the town in the Garrison Keillor short stories where "all the children are above average." Such systems fail to discriminate among good and bad service providers, and researchers have confirmed that there is often no real relationship between rating and quality. There is no evidence that an Uber driver with a rating of 4.9 is better than one with a rating of 4.6, even though the latter is in danger of being kicked off the Uber platform.

One underlying reason for the Lake Wobegon effect is that when we are unhappy at an interaction, many people follow the maxim "if you can't say anything nice, say nothing at all." Some leave no rating after a substandard Airbnb visit because they do not want the awkwardness that may go along with putting out a negative evaluation for the world to see. A reputation system acts as a guestbook at a bed-and-breakfast or a small museum: we leave comments and it looks nice, but it does not solve the hard problems of establishing trust. On eBay, a pioneer in online reputation, 99 percent of ratings are positive, even though one investigation put the number of dissatisfied participants at about 20 percent. When people rate each other on reputation systems most are generally being polite, not rendering judgment. This is perfectly appropriate behavior, but it makes the reputation systems useless.

The distorted rating distributions serve the interests of the platform owners, making the platform appear to be a higher-trust environment than it really is. Because of missing reviews and our tendency to be polite, reputation systems hide the level of dissatisfaction on a platform.

A second problem with reputation systems is that, even if most ratings are positive, providers live in fear of the occasional bad review that pushes them down in the search results or gets them removed from the platform entirely, depriving them of their livelihood. Service providers on a platform with a reputation system live in a Panopticon—always being watched, always being assessed. It's like living in an environment covered by unreliable CCTV cameras, which record images that may or may not reflect reality. Faced with the threat of a bad review, some service providers engage in compliant, indulgent, "emotional labor," catering to the whims of their most entitled customers. Drivers may or may not be good drivers, but they probably will not show it if they are in a bad mood. Bad reviews often have little to do with an objective evaluation; such systems have been shown to reflect and hence perpetuate patterns of prejudice among those doing the ratings.

Are there ways to tweak rating systems so they work better? It's a tempting proposition, but I think it is a dead-end. Most of us avoid giving bad ratings for good reasons: mutual assessment and reporting is a snitch system, incompatible with friendly and collaborative peer-to-peer relationships. It's a set of behaviors that belongs in a police state, and which has little place in an open and democratic society.

So how else could we deal with trust on platforms? While there are no simple and reliable answers, there are many sources of inspiration inside and outside the world of technology. Here are a few.

Some platforms avoid the problem completely. Craigslist and Kijiji make no claim to vouch for the parties in an exchange, and they expect buyers to beware. Like other listing services, they avoid getting involved in the actual transaction between buyer and seller, and take only a small fee from advertisers as their income. Stocksy United is another site that does not have a big need for a trust system because there is little room for deception; a photograph is what it is, and can be displayed and seen on the site before purchase.

Other successful technology-oriented communities have adopted a mixture of approaches. Internally, personal recommendations or one-on-one mentoring can play an important role, as in the Debian community that produces a leading Linux distribution. Personal invitations can also help to filter out unwanted members, as in the arXiv community that maintains an important pool of scientific papers and working papers. Wikipedia has its messy hierarchies and occasional lapses into fiefdoms, but remains remarkable for all that—or maybe because of all that—and its graduation of articles where trust is needed ("controversial places") helps to limit the need for formality.

Platform cooperatives share a problem with VC-funded platforms: the platform owner has an incentive to make the platform community appear to be working well, and to downplay or hide problems that arise. Any trust system needs to be externally auditable to have any credibility. We know very little about how (or if) Airbnb or Uber reputation systems really work, because they are hidden. The incentives to cover up failures become particularly strong when fortunes depend on a successful IPO. If there is not a significant tension between an independent auditor and system owner, then the system is probably not doing its job.

Unlike VC-funded platforms, platform cooperatives should accept independent external audits of what actually happens on the platform. The experience of fair-trade activists provides a source of inspiration:

consumers pay a premium for products that support a fair supply chain, and economists Kimberly Ann Elliott and Richard Freeman have looked into the fair trade certification systems that seek to establish trust (how do we know the producers are following through on their claims?). They went into their study expecting that the most effective system would feature one well-recognized certification system, and were surprised to find that an ecosystem of competing certificate programs had many benefits.

There is a need for similar independent certification programs around platform cooperative offerings. The best forms for such programs will be discovered by experimentation, and here cooperatives have a real advantage. Venture-funded platforms are impelled to deliver a successful IPO exit for their investors, and so have too much to lose from a bad report. Maybe platform cooperatives can avoid the closed and secretive character of those companies, and experiment in the open.

28. WHY PLATFORM CO-OPS SHOULD BE OPEN CO-OPS

MICHEL BAUWENS AND VASILIS KOSTAKIS

"What if this was no longer capitalism, but something worse?" McKenzie Wark's statement, which opens his chapter in this book, eloquently summarizes the growing criticism of profit-maximizing business models within the so-called collaborative sharing economy. That "something worse" appears to take the form of a new kind of feudalism. If feudalism was based on the ownership of land by an elite, the resource now controlled by a small minority is networked data. We cannot, therefore, be content with cooperative alternatives designed to counter mere capitalism.

Commons-based peer production, a term coined by Yochai Benkler, has brought about a new logic of collaboration between networks of people who freely organize around a common goal using shared resources, and market-oriented entities that add value on top of or alongside them. Prominent cases of commons-based peer production, such as the free and open-source software and Wikipedia, inaugurate a new model of value creation, different from both markets and firms. The creative energy of autonomous individuals, organized in distributed networks, produces meaningful projects, largely without traditional hierarchical organization or, quite often, financial compensation.

This represents both challenges and opportunities for traditional models of cooperativism, which date back to the nineteenth century, and which have often tended to gradually adopt competitive mentalities. In general, cooperatives are not creating, protecting, or producing commons, and they usually function under the patent and copyright system. Further, they may tend to self-enclose around their local or national membership. As a result, the global arena is left open to be dominated by large corporations. Arguably, these characteristics have to be changed, and they can be changed today.

The concept of open cooperativism has been conceived as an effort to infuse cooperatives with the basic principles of commons-based peer production. Pat Conaty and David Bollier have called for "a new sort of synthesis or synergy between the emerging peer production and commons movement on the one hand, and growing, innovative elements of the cooperative and solidarity economy movements on the other." To a greater degree than traditional cooperatives, open cooperatives are statutorily oriented toward the common good. This could be understood as extending, not replacing, the seventh cooperative principle of concern for community. For instance, open cooperatives internalize negative externalities; adopt multi-stakeholder governance models; contribute to the creation of immaterial and material commons; and are socially and politically organized around global concerns, even if they produce locally.

We will outline a list of six interrelated strategies for postcorporate entrepreneurial coalitions and a mode of value creation that is autonomous, fair, and sustainable. The aim is to go beyond the classical corporate paradigm, and its extractive profit-maximizing practices, toward the establishment of open cooperatives that cultivate a commons-oriented, ethical economy.

First, it's important to recognize that closed business models are based on artificial scarcity. Though knowledge can be shared easily and at very low marginal cost when it is in digital form, closed firms use artificial scarcity to extract rents from the creation or use of digitized knowledge. Through legal repression or technological sabotage, naturally shareable goods are made artificially scarce so that extra profits may be generated. This is particularly galling in the context of life-saving medicines or planet-regenerating technological knowledge. Open cooperatives, in comparison, recognize natural abundance and refuse to generate revenue by making abundant resources artificially scarce.

Second, a typical commons-based peer production project involves various distributed tasks, to which individuals can freely contribute. For instance, in open software projects, participants contribute code, create designs, maintain the websites, translate text, co-develop the marketing strategy, and offer support to users. Salaries based on a fixed job description may not be the most appropriate way to reward those that contribute to such processes. Open co-ops, therefore, practice open-value accounting or contributory accounting. The Sensorica project, which produces scientific instruments, expects contributors to log their contributions and, after peer evaluation, they are assigned a certain amount of "karma points." Any income the contributions generate then flow to contributors according to the points they accrued. This model is an antidote to the tendency in many firms for just a few well-placed contributors to capture the value that has been co-created by a much larger community.

Third, open cooperatives can secure fair distribution and bene-fit-sharing of commonly created value through "CopyFair" licenses. Existing copyleft licenses—such as Creative Commons and the GNU General Public License—allow anyone to reuse the necessary knowledge commons on the condition that changes and improvements are added to that same commons. That framework, however, fails to encourage reci-procity for commercial use of the commons, or to foster a level playing field for ethical enterprises. These shortcomings can be met through CopyFair licenses that allow for sharing while also expecting reciprocity. For example, the FairShares Association uses a Creative Commons non-commercial license for the general public, but allows members of its organization to use the content commercially.

Fourth, open cooperatives are able to make use of open designs to produce sustainable goods and services. For-profit enterprises often aim to achieve planned obsolescence in products that would wear out pre-maturely. In that way, they would maintain tension between supply and demand and maximize their profits; obsolescence is a feature, not a bug. In contrast, open design communities, such as these of the Wikispeed car, the Wikihouse, and the RepRap 3D printer, do not have the same incentives, so the practice of planned obsolescence is alien to them.

Fifth, and relatedly, open cooperatives reduce waste. The lack of transparency and penchant for antagonism among closed enterprises means they will have a hard time creating a circular economy—one in which the output of one production process is used as an input for another. But open cooperatives can create ecosystems of collaboration through open supply chains. These chains can enhance the transparency of the production processes and enable participants to adapt their behavior based on the knowledge available in the network. There is no need for overproduction once the realities of the network become common knowledge. Open cooperatives can then move beyond an exclusive reliance on imperfect market price signals and toward mutual coordination of production, thanks to the combination of open supply chains and open-book accounting.

Sixth, open cooperatives can mutualize not only digital infrastructures but also physical ones. The misnamed "sharing economy" of Airbnb and Uber, despite all the justified critique it receives, illustrates the potential in matching idle resources with demand. Co-working, skill-sharing, and ride-sharing are examples of the many ways in which we can reuse and share resources. With co-ownership and co-governance, a genuine sharing economy could achieve considerable advances in more efficient resource use, especially with the aid of shared data facilities and manufacturing tools.

How, then, does the concept of platform cooperativism relate to the notion of open cooperativism? Cooperative ownership of platforms can begin to reorient the platform economy around a commons-oriented model. We highlighted six practices that are already emerging in various forms but need to be more universally integrated. We believe that a chief ambition of fostering a more commons-centric economy is to recapture surplus value, which is now feeding speculative capital, and re-invest it in the development of open, ethical productive communities. Otherwise, the potential of commons-based peer production will remain underdeveloped and subservient to the dominant system. Platform cooperatives must not merely replicate false scarcities and unnecessary waste; they must become open.

PART 4

CONDITIONS OF POSSIBILITY

SHOWCASE 2 : ECOSYSTEM

The "killer apps" of platform capitalism didn't come out of nowhere. The big companies that rule the Internet aren't coming to dominate just because of a good idea and a charismatic founder; they grow out of supportive ecosystems, including investors, lawyers, sympathetic governments, and tech schools. Perhaps most important is their culture—the festivals, the meetups, the memes, the manifestos—that share norms for what kinds of practices are expected and celebrated. To change these norms, we need to cultivate an ecosystem for platform cooperativism. These projects demonstrate that this effort is already under way.

Project Name: Loomio Cooperative Ltd.
Completed by: Mary Jo Kaplan
Location: Wellington, New Zealand, and Providence, Rhode Island
URL: loomio.org

Loomio is a worker-owned cooperative that is building a tool for collaborative decision-making used by thousands of cooperatives, community organizations, social movements, and government initiatives across the globe. Loomio enables people to contribute to decisions that affect them, to drive self-determination, better decisions, stronger communities, and engaged workplaces. Loomio's users are incredibly diverse and so are the ways they use the tool, from day-to-day operational decisions in companies, to collaborative policy development in government, and community engagement by NGOs. In late 2015 we released Loomio 1.0, a mobile-first interface with a focus on interoperability and an automated subscription system designed for growth.

Loomio is a robust social enterprise with an ethical business model that was started four years ago. As a worker-owned cooperative, Loomio is owned collectively by the people building it. The current board of directors is made up of four members and one former member. As an open source tool and global social enterprise, we actively engage developers, contractors, activists, investors, customers, advisors, and other stakeholders to work with us to make a better product and company.

We've attracted talented people to work well below market rates without issuing traditional equity. We bootstrapped for four years through consulting revenue, loans, crowdfunding, grants, and donations. Also, we've attracted extremely valuable advisors by being genuinely mission-driven. In November 2015 we raised $450,000, using redeemable preference shares as an investment instrument that aligns with our social mission and cooperative structure while providing a fair return to investors.

Loomio is poised for growth in the second half of 2016 and beyond. Our vision is for Loomio to be a ubiquitous technology, seamlessly integrated with other tools people use every day. By growing revenues based on providing value to our customers, not selling user data or advertising, our success will be based on serving customers' needs and realizing workers' collective values and commitment to social impact.

Project Name: The FairShares Model
Completed by: Rory Ridley-Duff
Location: Sheffield, England
URL: fairshares.coop

The FairShares Model is a suite of intellectual properties developed during social enterprise research programs at Sheffield Business School in order to support the creation and development of solidarity cooperatives. The working assumption is that the exclusion of primary stakeholders from enterprise ownership and governance harms the well-being of members and their host community. The association's model rules encourage four classes of membership: founders, labor, users, and investors.

FairShares IP is owned by the member(s) who contribute to it. Members license it to the FairShares Association for use in charitable and commercial projects. If members leave, both they and the association retain non-exclusive licences. Policy decisions are taken by members. Decisions on marketing are usually taken by members and supporters in a Community Forum (on Loomio.org). Members communicate with supporters and each other via MailChimp, Facebook, and Loomio.

FairShares IP is freely available in PDF format from fairshares .coop (with editable versions that can be supplied directly by email or through a shared Dropbox). Supporters can join a MailChimp list, make regular financial contributions, and join the Community Forum at fairshares-association.com. Some help is provided free at the point of use or via Loomio discussion boards, and some members offer paid consultancy services.

A 2015 survey showed that FairShares IP is being actively used in thirteen countries including the United Kingdom, the United States, Australia, and New Zealand, and has global reach through its inclusion in a social enterprise textbook. Clients in the United Kingdom, the United States, and Ireland created FairShares enterprises in 2015, and we are supporting new projects in Australia and Croatia. We currently have 1,629 subscribers on MailChimp, 885 followers on Facebook, 69 supporters in our Community Forum, plus 11 labor, 6 user, and 4 founder members.

Project Name: Swarm Alliance
Completed by: Joel Dietz
Location: Palo Alto, California
URL: swarmalliance.com

Swarm Alliance is a network of aligned organizations that are dedicated to creating a world of abundance. We've been especially involved in the world of collaborative governance and the commons. Most of the technology developed through our network has involved blockchain in some fashion.

The Swarm Alliance uses the Distributed Collaborative Organization model that we co-invented at a legal conference at Harvard University. This model was designed to have two levels, one for rapid decision making (delegates) and another for approval of large decisions (members). The Swarm Alliance currently has three delegates and approximately one thousand members. It is the first and largest operational example of an organization hosted entirely on a blockchain.

We started with an approximately $1 million crowdfunding campaign around our own blockchain-issued asset, the Swarm coin. After much of the funds were exhausted in the process of our own legal research and the coin's price fluctuations, we financed development through corporate partnerships. At this point we are re-engineering our initial proof of concepts (developed on Bitcoin) to release ready concepts on Ethereum.

We are currently building a global ambassador network through training events in various cities across the globe. We are also currently exploring other projects that might bridge to a mainstream audience and serve as a proof of concepts for both the future of governance and community abundance. As with our original concept, we expect new forms of crowdfunding to have a major role in this, especially around blockchain-hosted organizations.

Project Name: Ms., The Madeline System
Completed by: Eden Schulz and Brendan Martin
Location: New York City
URL: http://theworkingworld.org/us

The Madeline System networks local, cooperative investment funds, giving them the ability to stay community-controlled and yet gain the benefits of scale. Through pooling the investments of a network of funds, Madeline dramatically increases the market leverage of each fund and allows it to command far better investment terms. By joining the funds into a network of mutual accountability and sharing, Madeline also brings high-quality business assistance to even the smallest partners. The result is a distributed, non-extractive financial system with communities in the position of power.

Organizations using Madeline become members and own their assets together cooperatively. Decisions are made democratically, with the governance structure designed to minimize central control and maximize local autonomy.

The initial development of Madeline was made possible by a generous donation. As members become operational, they share the cost of maintaining Madeline using income from their local investments. Cooperation is the heart of Madeline, and it is the only way that the scaled benefits of the system can be brought to small loan funds at a reasonable price-point. Cooperation is also the key to bringing capital to small funds at terms patient and low enough to allow the nurturing of local businesses.

In the past year, we have brought together our first cohort of member funds who will form the foundation of the Madeline user-member base. Over many years, we have successfully built and used a prototype of the Madeline platform, and we have designed and begun building the launch version of the system. We expect to begin using this version with our first members before the end of 2016. By the end of 2017, we project a robust network of members using the system across the country.

Project Name: Purpose Fund
Completed by: Alexander Kühl and Armin Steuernagel
Location: Berlin, Germany
URL: http://purpose.ag

Purpose Fund is a startup fund for an ecosystem of purpose-driven entrepreneurs. The fund invests in companies that have a "purpose ownership" form, which means that the company owns itself.

There are two basic principles of purpose ownership: (1) voting rights are decoupled from dividend rights and are held by those who lead the company or are actively involved; (2) profits are a means to an end and not an end in themselves. In practice this means that dividend rights are held by investors without voting rights. Dividends are capped and profits are reinvested or used to pay back investors. We call this the self-determination of the company. Because of this ownership structure, decisions are not driven by shareholder value maximization, and the company is not an object of speculation.

Purpose ownership—which can take different legal forms, including cooperatives—is a clear signal to employees, investors, clients, and other potential collaborators that their contribution benefits the purpose of the company rather than the private wealth of equity holders. This enables close cooperation between the startups because they know that the value that is thus created cannot be privately extracted. Purpose companies voluntarily share employees and code, creating benefits similar to those found in large corporations while retaining decentralized ownership.

The fund invests in platform tech companies that are working on the future of work, insurance, the sharing economy, decentralized internet, and open data. It leaves voting rights with entrepreneurs and its investment strategy is evergreen rather than exit-driven. Rooted Internet, itself a self-owned company partly controlled by its startups, raises money by selling non-voting shares. It promises never to extract more money out of the startup ecosystem than a certain capped dividend.

Project Name: rCredits
Completed by: William Spademan, Executive Director of Common Good Finance
Location: Western Massachusetts
URL: rCredits.org

The rCredits® system is a complete alternative banking system for the common good, using a local credit/debit rCard® with incentive rewards and no fees for buyers or sellers. This system empowers any community to decide together what's best for them, with money to fund it, using an innovative participatory decision process.

Anyone can open an rCredits account with a member invitation. Transactions, decisions, and funding are managed independently and transparently within each region or community. Every member gets one vote and can vote directly on any substantive issue.

About three person-years have gone into developing a smartphone app and backend server. In addition to that time and effort, hundreds of private donations provided about $300,000 to cover thirteen years of work. We built a self-funding mechanism into the design, so as the rCredits system grows, it becomes more and more sustainable. For the past year, we have been preparing for growth by automating administration of the system and segmenting it by community, so each community can manage its own affairs. The system includes protocols for essential cooperation and mutual oversight between rCredits communities.

We now have active rCredits pilot projects in western Massachusetts, southeastern Vermont, and Ann Arbor, Michigan. We are inviting interested individuals in other communities to begin promoting rCredits locally. In April we expect to release a revised rCredits app that works on both iPhone and Android. The rCredits system has handled three-quarters of a million dollars in transactions, and we have freed up tens of thousands of dollars for our participating communities to use for grants, incentives, and investments in sustainability. Our top achievement is the nearly seamless integration with the mainstream economy, allowing the system to grow smoothly in order to provide a democratic, community-centered replacement for our unjust and destructive global economic system.

Project Name: External Revenue Service
Completed by: Max Dana
Location: Brooklyn, New York
URL: externalrevenue.us

The External Revenue Service (ERS) is a peer-to-peer tax system designed to make it easier for people to share their disposable income with the things they care about most. The rules of the ERS are simple. Givers pledge a percentage of their income to be automatically redistributed to a portfolio of receivers each month, and receivers must make a giving pledge of their own in order to receive the funds that have been pledged to them. In the ERS, everyone is a philanthropist.

The ERS is a distributed collaborative organization and is not owned by anyone. It is a network of contributors and users invested in the maintenance and development of the system. We currently collaborate via an open Slack team that is modeled on similar Slack teams piloted by OpenBazaar and Backfeed.

We have had the good fortune of having some very smart people come forward to help define and refine the vision of the organization. As we transition from concept to product, we are actively seeking more developers to contribute to the codebase as well as legal experts to help us navigate the complex regulatory landscape (particularly with respect to digital currencies). Financially, the organization is bootstrapped with a small pool of funding from the community and is committed to forgoing traditional investment in favor of voluntary funding from the network.

In May 2015, the ERS was just an idea born out of frustration with fundraising and income inequality, but under the mentorship of Gary Chou at Orbital NYC we were able to validate some early assumptions and raise money via Kickstarter to bring together a seed community for the Weird Economics Summit in NYC in November 2015. In 2016, we hope to develop a minimum viable product of the platform and pilot it in partnership with like-minded organizations.

Project Name: Data Commons Cooperative
Completed by: Noemi Giszpenc
Location: Massachusetts
URL: datacommons.coop

The Data Commons Cooperative brings together cooperative, solidarity, social, generative, "new" economy organizations that are cataloging or mapping some slice of that space. Many organizations want to publicize the existence of alternatives and enhance the connections among them; the data-sharing cooperative makes it easier for members to gather, share, maintain, display, and deploy information about the economy they care about.

The DCC is owned by its data-sharing members. It is incorporated in the state of Massachusetts and has an elected board of directors. The bylaws are at member.datacommons.coop/bylaws. Decisions are consensus or super-majority. The membership ratifies yearly budget and capital plans. The board of directors receives membership applications and votes on accepting new members. Each member chooses what data to share and how or with whom. When we have permanent staff, we'll have a democratic workplace.

Making it work is a struggle. This feels like necessary infrastructure, but that's about as sexy as paving a road. We've had devoted tech volunteers, and received support from foundations, donors, and the government—but, it hasn't been enough to cross the threshold into sustainable operations. Our current plan is to focus on fundraising and cross-subsidization from commercially viable products.

We have thirty members, an elected board of five representatives, a user-friendly search-and-display codebase (Stone Soup), and the beginnings of a technical solution to the directory-updates challenge (daff). We've been featured in Grassroots Economic Organizing, Shareable, and Community-Wealth.org; helped shape a map of the solidarity economy in the United States (solidarityeconomy.us); and will play a role in shaping a census of co-ops in the Northeast. In the next few years, we hope to raise enough money to pay staff and provide more tailored member services.

Project Name: Coliga
Completed by: Pedro Jardim
Location: Berlin, Germany
URL: coliga.co

Coliga is a platform that helps any network create a vertical marketplace and become more self-organized and financially sustainable. We make it easy for networks, like coworking spaces and cooperatives, to share in any revenue generated from jobs and connections they facilitate for their members.

Many networks have strong local and global brands, and they regularly receive job requests and offers for their members. Coliga improves how networks capture and channel these opportunities to their communities, with tools to find the best people for the job and split the value with them once the job is complete. That way value is created and shared within the network, rather than being taken away.

We've structured the ownership of our company in a way that maximizes our long-term social impact. We've separated voting shares (allocated at nominal value to company managers) from income-paying shares, which means investors get dividends but not the right to influence the business strategy in a way that sacrifices purpose in favor of short-term profits. Our bylaws also dictate that all profits are to be reinvested into the company.

Coliga is the latest project of people behind Apoio, a self-organized community cleaning service, and Agora Collective, a leading creative space in Berlin. We're part of the OuiShare community and won a 2015 OuiShare award in the category of collaborative economy. We also belong to Rooted Internet, which invests in purpose-driven companies. In the coming years we are going to build a diverse global community of like-minded peers to enable new forms of collaboration and resource sharing among networks.

Project Name: CommunityOS: Callicoon Project
Completed by: Ashley Taylor
Location: Bushwick, Brooklyn, New York
URL: futureculture.how/community-os-callicoon-project

CommunityOS is a layer in the blockchain operating system. It is a community network for exchanging resources, creating added value, and developing cooperatives using complex barter systems and reputation calculators. It will first be used at scale in the agricultural community of Sullivan County, New York, for the Callicoon Project, and for event communities in New York City.

The project collaborates closely with ConsenSys, a blockchain company developing tools that allow cooperatives to easily form and manage their resources, make decisions, collaborate, transparently distribute equity and shares, and evolve themselves. These include boardroom.to (governance and decision making), inflekt.us (community network and events), weifund.io (decentralized crowdfunding and equity distribution), and uPort.me (cryptographic identity and reputation container). The Inflekt platform serves as the interface to CommunityOS.

On January 16, 2016, we had a potluck event with interested people in the Callicoon community at one of the well-known farms. We introduced the ideas of cooperative management and resource banks. We are currently developing a cooperative model for a red-meat processing facility. It will include day-to-day management, community investment, and community insurance, all contractually secured on the blockchain.

We are also building a resource bank that will allow communities to identify, search, exchange, barter, and gift between each other. Our goal is to develop new metrics for value and new marketplaces to recognize that value for each community, creating a diverse ecosystem of community-based support and connection.

We are currently supported by ConsenSys and simultaneously applying for grants for the research on the cooperative economic models that will be prototyped alongside the implementation of CommunityOS in context, specifically in Callicoon.

Project Name: Backfeed
Completed by: Julian
Location: Tel Aviv, Israel
URL: backfeed.cc

Backfeed is an engine for decentralized collaboration. It builds upon the power of open-source collaboration and enhances it with a distributed governance system for decentralized value production and distribution. We've developed an algorithm that allows large groups of users to contribute freely and spontaneously to an enterprise, determine the perceived value of each contribution, and allocate influence and rewards accordingly. Backfeed makes it possible for decentralized networks of peers to coordinate themselves indirectly in order to achieve the full potential of collective intelligence.

Backfeed is a privately held company founded by Matan Field and Primavera de Filippi.

Backfeed is currently funded exclusively by angel investments. We are now raising funds in order to support our operations. Later this year, we intend to launch a crowd-sale of Backfeed tokens that will serve to fuel the Backfeed protocol and platform.

We are already at an advanced stage of development of our core technology—the Backfeed protocol engine—which provides an outstanding breakthrough in the field of decentralized collaboration. We've built an API layer that lets others connect to and operate our protocol, which is currently used by our partners at Slant News to enable collaborative editing for articles on their platform. There are various other companies and organizations interested in using our system, and we're currently establishing partnerships and collaboration opportunities with strategic partners and like-minded individuals. In parallel to that, we are also developing our own app, as a testing ground for experiments and for the development of further components of the protocol stack.

Project Name: My User Agreement
Completed by: Anna Bernasek, D.T. Mongan
Location: New York City
URL: myuseragreement.com

Myuseragreement LLC offers individuals terms and conditions intended to protect their legal rights in the data they create. For the first time consumers have a real option; it's no longer necessary to blindly accept unfair terms from service providers. By adopting myuseragreement, consumers can protect their rights in just a few easy steps. Individuals sign up by submitting their email, and when we have reached a statistically significant number of sign-ups we will go live. At that point, we send each user an email with a link that allows installation of a simple digital marker alerting anyone interacting with a user's device to your terms and conditions.

Myuseragreement was founded as a public service in 2015 by Anna Bernasek and D.T. Mongan, the authors of *All You Can Pay: How Companies Use Our Data to Empty Our Wallets* (Nation Books, 2015). Anna Bernasek is a journalist and author based in New York. Among other publications, she has written for *The New York Times*, *Newsweek*, *Fortune*, *Time*, *The Australian Financial Review*, and *The Sydney Morning Herald*. D.T. Mongan is a lawyer based in New York. He wrote the terms and conditions for myuseragreement. Those terms and conditions are laid out on our website in a simple and straightforward way.

Myusergreement is in the sign-up phase and our focus is on spreading the word about this new Web tool for consumers. We welcome any interest and suggestions from the public. Please be in touch through the website or with the founders directly. We believe that by acting together we can develop a better and safer Internet.

29. BEYOND LUXURY COOPERATIVISM

JOHN DUDA

The long neoliberal revolution in the psychology of advertising made it possible to mistake the atomized actions of consumers for commitment and authenticity: cool becomes commodity, and the production of the individual consumer replaces the project of collective liberation. This process of recuperation—in which individuals realize themselves primarily in and through the choices they make in the marketplace—can also infiltrate our visions of alternatives, with our utopian imagination all too often already colonized by its own undoing. The success or failure of platform cooperativism may lie in whether it can escape becoming absorbed into the consumerist colony.

Neoliberalism's cultural revolution arguably began in the counterculture of the 1960s, in which opposition to systemic injustice and alienation found expression in new patterns of life. But choosing to live differently could all too easily mean prioritizing a new form of "empowered" consumer identity over collective political engagement. It's no coincidence that this historical moment gave rise, on the one hand, to the early visions of a networked economy, and on the other, to the first significant wave of cooperative development in the post-WWII period. In both, the individual, reduced to a consumer, stands in for the collective subject of political action, and alternatives become spaces of withdrawal, not engagement.

Consider the expansive vision of the *Whole Earth Catalog*, inviting the people at the intersection of the thriving counterculture and a nascent cyberculture to take up the "tools" they will need to rebuild

Spaceship Earth. The project of liberation, in the *Catalog*, is quite literally a shopping exercise: one picks out the ideas and technologies that construct and confirm a new identity in (nominal) opposition to the mainstream—hence the mélange of yurts and primitive computers, cybernetics, and new age theology that floats across its pages. This early cyberculture shares more than we might expect with the closed corporate networks of the 1980s—in an infamous CompuServe ad, we see the way utopian promise feeds on (literally white) fear of urban space—a retreat from political community into networked consumer identity. Today's app-mediated landscape, despite some innovations in network topology, and a new relationship between economic and racial privilege and urban geography, offers another iteration of the same: digital platforms allow us to pick and choose the communities we connect with and commit to, replacing the messy work of political solidarity with the frictionlessness of "disruption."

The collective and cooperative workplaces emerging from the late-1960s counterculture followed the same neoliberal logic. These alternative institutions were an escape route—a way for those with the requisite privilege to construct bubbles of autonomy, outside the alienating corporate workplace and market. The most prolific and enduring cooperatives of the era were the various food cooperatives that sprung up across the country, quite literally built on the idea of aligning consumer choices with new values.

It isn't that cooperative consumer purchasing can never result in powerful or inclusive collective institutions—Japan's Seikatsu Cooperative, built largely by women and an influential force in food policy, shows otherwise. But where Seikatsu is premised around rejecting "choice" as illusory in favor of a limited set of sustainably farmed staples, the food co-ops of the 1960s (or at least the ones that were viable financially) promoted a mode of opposition to the status quo mediated by consumer choice. Taking stock of this wave of cooperatives in 1979, David Moberg wrote, "Many alternative institutions were rendered relatively harmless as another market choice." Much as the counterculture technologists unwittingly prepared the way for

183

the corporate Internet, the food cooperative movement largely fueled the commodification of natural and organic food as an upscale market segment, not the development of a just and sustainable food system.

The budding movement for platform cooperatives should avoid recapitulating this trajectory. One place to look for guidance can and should be the current wave of cooperative development, which looks very different from its 1960s and '70s counterpart, especially when it comes to worker cooperative development. While the worker cooperative sector remains quite small, the movement to expand it is robust, dynamic, and incredibly instructive when it comes to thinking about building a more inclusive economy.

For one thing, the contemporary worker cooperative movement is a lot less white than most people imagine. The largest worker cooperative in the country is the two-thousand-person Cooperative Home Care Associates in the Bronx, whose worker-owners are overwhelmingly women of color. Strong networks of interlinked worker cooperatives are developing in historically marginalized communities of color across the country—like the house cleaning and food processing cooperatives incubated by Prospera in Oakland, or the Evergreen Cooperatives in Cleveland, with worker-owners running a multi-acre urban greenhouse, a green energy company, and a large-scale commercial laundry. Such developments did not arise automatically, but resulted from the careful work of organizers, community advocates, and nonprofit developers who have started from the premise that the point of building worker cooperatives is first and foremost to create an economy owned by the people who have been traditionally locked out or pushed to the side—immigrants, the poor, the formerly incarcerated, and other victims of business as usual.

This kind of shift in focus means leaving behind some of the cherished emphasis that the last cooperative wave put on autonomy and independence; instead, cooperative developers have been enthusiastically exploring ways to partner with city governments, labor unions, forward-thinking philanthropy, and impact investors to create mechanisms to finance and support the work of building a more democratic economy. Cooperative purity can easily become an obstacle to achieving meaningful scale and inclusive impact. For instance, insisting that members self-finance their own enterprise risks creating a closed circle of relative economic privilege rather than a movement to truly democratize capital. Similarly, insisting on autonomy from government intervention and support means that the policies behind traditional economic development will continue to grind away, subsidizing corporations and leaving cooperatives to fend for themselves. The point, as cooperative advocates have come to realize, is not to create maximally pure alternatives that help a few escape an alienating system, it's to take seriously a long-term project of actually transforming what counts as business as usual—in the direction of equity, democracy, solidarity, and sustainability.

What can a movement for platform cooperativism learn from all this, then? Perhaps the most crucial lesson is to understand why a deep commitment to inclusion needs to be a grounding principle, not an afterthought. In the absence of institutional designs and development processes that put the needs and aspirations of the marginalized first, platform cooperatives might very well create new spaces of exciting digital freedom for their members, but in so doing only put a new spin on the existing patterns of inequity.

Part of this involves tackling the question of financing. Worker-cooperative developers have been very busy identifying pathways to scale their efforts by making public, philanthropic, and impact capital available to new and existing democratic workplaces. Platform cooperatives looking to reach a scale comparable to that of the dominant platforms are going to need to follow the lead of the worker cooperative movement and get creative and intentional about their financing strategies.

Similarly, there needs to be a recognition among platform co-op developers that no matter how good your intentions are or how technically sophisticated your project might be, market dynamics alone tend toward inequitable outcomes. Changing this equation means finding ways to connect cooperative initiatives to resources that aren't completely constrained by the logic of profit. The Evergreen Cooperatives, for instance, leverage the procurement dollars of large place-based nonprofits like universities and hospitals to create a partial shelter from the market—building the foundations of a long-term effort to create worker-owned jobs in severely distressed communities.

Above all, it's imperative that people drawn to platform cooperativism realize that the smooth space of disruption is not where real collective power can be built. No matter how shiny its apps or how ostensibly democratic its process, an institution built on a foundation of individual privilege to opt in will do nothing to challenge and transform the underlying dynamics of our economic system. Instead, we need to look past the mirage of superficial alternatives and get busy building real relationships across sectors and throughout our communities. This means committing to investments in deep collective education, long-haul transformative organizing, and intentional efforts to shift resources to and building power in the communities that today's economy locks out. While technology might help facilitate this work, it cannot magically replace it.

30. MONEY IS THE ROOT OF ALL PLATFORMS

BRENDAN MARTIN

To use the term "platform cooperativism" is to sum up an incredible swath of economic theater in just two words. The term, first, presumes the concept of a platform as a common infrastructure people use. But putting the word "cooperativism" after it is to suggest the central question of a platform is who owns it: is it owned cooperatively by all those who use it, or by just a few who extract value from the rest? I contend that, despite the great advances in technology that have made Internet-based platforms explode into our lives, the contest over who owns our platforms is not at all new. It is, rather, one of the central struggles of civilization, from the first wars over the Fertile Crescent to the present battles about Uber.

Money, or finance, is not just what people use to buy platforms; it *itself* is a platform. Finance is the infrastructure of exchange as old as history, and the battle for its control has been raging since the first coins were minted and the first debts recorded. Finance is so fundamental to our economy that we can easily miss it, as a fish may not be aware of the water it swims in, but we cannot afford to do this. Finance is not just one of the oldest platforms, it is perhaps the root that so many other platforms are controlled by. This means that challenging the structures of finance may be the key to challenging the ownership of any platform. Recent advances in platform technologies offer new tools for doing just that.

To put into perspective the role that finance plays in the ownership of other platforms, consider what keeps any of us from making our own Uber or some other large platform. Our attention is focused on the breakthroughs in technology, like smartphones and the Internet, that have made these platforms possible. But technology is not the barrier to making a platform; if anything, the advances in technology have made platform-building easier. The barrier is finance. How else, but with mountains of money, could a few unelected men—as most platform owners are—command hundreds of programmers and thousands of marketers and lawyers to build a platform they alone own and which millions depend on? It is the platform of finance, and the intensely unequal control of that platform in our world today, upon which so many other unequal platforms have been built.

Privately held money is not the only way to build a platform, however. Platforms have been built through government—such as a highway system, or an electrical grid, or the Internet itself—and these are generally controlled (in theory at least) not by investors but by citizens. Platforms can also arise from the spontaneous efforts of people organized by unions, communities, or movements, such as farmers' producer cooperatives. But, by and large, the new platforms suddenly impacting our lives are mainly built on the platform of private finance.

BREAK THE BANK AND LIBERATE THE PLATFORM

It might be tempting to conclude that, given the domination of private finance in society as a whole, building a cooperative platform would require something like a revolution. But this is not true, and every cooperative in the world proves this every day.

Consider a worker-owned factory. In the "standard" investor-owned factory, finance gets pooled together; it buys the labor and materials to build and run the factory, it hires and fires workers, and it keeps whatever profit comes in. In a worker cooperative, however, the workers gather together; they borrow the money to build and operate the factory, they pay it back when they don't need it, and they keep whatever profit comes in. This worker cooperative turns the platform of finance on its head. It was in Mondragon, the great mecca of cooperatives in northern Spain, that the workers declared that capital must be subordinate to labor to allow cooperatives to flourish. We must do the same if we want to build platform cooperatives. No plan to build one will make sense unless control over finance is a key part of its strategy.

The logical conclusion of this is to turn the platform of finance itself into a cooperative. Then, each new cooperative enterprise would not turn into a struggle to resist centralized finance, but cooperatives could be built repeatedly on their shared platform of cooperative finance.

Mondragon built its cooperative financial platform the old-fashioned way: the members made a brick-and-mortar bank, the Caja Laboral, put all their personal and corporate savings into it, and built it into a financial powerhouse over decades. Upon this, they built an extraordinarily expansive and vibrant cooperative economy that has impacted their entire region and turned a tiny mountain village into the headquarters for many multi-billion-dollar worker-owned businesses.

We can follow Mondragon's example and build institutions that directly compete with centrally owned finance. This is what credit unions, public banks, and cooperative lenders like my organization, The Working World, have been doing for a long time. But the struggle to gain ground against the U.S. financial system is much harder than in the isolated village of Mondragon in the 1950s. So far, the promise of financial cooperativism for all remains a very uphill battle.

This is where the disruptive power of technology could become an opportunity. The Internet's capacity for enabling transactions among disparate people, for sharing information about products and investments, and for creating new currencies and decentralized means of exchange—each of these creates opportunities to subvert traditional finance. We have a window for taking advantage of them, and this is exactly what The Working World and our allies are trying to do. (For more about this effort, see our showcase on The Madeleine System in this book.)

THE GUERRILLA WAR OF PLATFORM COOPERATIVES

Cooperative businesses have managed to succeed in a world of centralized finance, and cooperative platforms can too, but it isn't always easy. Here are some lessons that we at The Working World have learned in our experience supporting cooperatives, both online and off.

1. Build on cooperative finance. Most cooperatives start with pools of money from people and communities. In this way, cooperative businesses carve out a bit of cooperative finance to build themselves on, and any would-be cooperative platform builder will have to do the same.

If you imagine getting financing for a platform, consider the relationship being offered to you by those who control the finance. Is it extractive? Will ownership, control, or rewards mostly go to the sources of money? What if things go at all wrong, and you can't meet your targets or make a payment—do you lose control? Resist the seduction of larger offers of money that come with extractive strings.

Look for cooperative lenders, credit unions, or CDFIs. Find impact investors, community investors, or crowdfunding opportunities. But don't assume these sources mean you don't have to pay attention; structure your deals to make sure the people most involved—the workers or users—are the owners and finance is the element that gets hired or fired.

2. Don't just build—convert. If you need to own a house, do you always need to build a new one? It is usually much easier to find an existing house and buy it—or, to put it another way, convert who owns it. The problem with many of our platforms is also simply who owns them, and we may find it is far easier to make cooperatives, including platform cooperatives, by converting existing businesses rather than building them from scratch.

It is predicted that a million profitable small businesses will close in the next decade because their baby-boomer owners are retiring. A growing body of entrepreneurs and policymakers is looking to save these business by selling them to their workers, thereby converting them to worker ownership.

There is no reason why we cannot consider conversion for online platforms as well, although many of them have become so overvalued that the financial barriers are considerable. It probably won't be possible to buy the biggest companies at the peak of their market value, but opportunities to change the equation exist. What if the users—aka, content-creators—of Facebook organized to demand ownership under threat of a strike? What if municipalities trying to fend off Uber allowed it to operate in exchange for some form of ownership for drivers? What if governments and organized groups of people acted in concert to depress a platform's market value to a price we would be able to pay? We could start by going after a platform most susceptible to this type of pressure, converting it to cooperative ownership, and making it an example for future conversions.

This is certainly not an easy strategy, but it might be easier than trying to build and finance a massive platform from scratch.

3. Organize. That brings me to my final recommendation in the guerrilla war against centralized platforms: organizing.

The practice of social-justice organizing is evolving these days. Rather than mostly combating businesses and demanding concessions, organizers and the organized are learning to create and control businesses. The work of organizing—surprisingly to some—develops many of the skills needed in a functioning business: communication, understanding the needs of others, and leading others toward common goals. Organizing is not the opposite of business, it is just the opposite of business for the benefit of a few. When attempting to build not just a cooperative business but a cooperative platform, the ability to organize large numbers of people becomes perhaps the central skill necessary for success.

In the effort to oppose centralized platforms, now is a time not just to fight, but to create. The most fundamental challenge may be to rebuild the underlying platform of finance itself.

31. FROM PEOPLE-CENTERED IDEAS TO PEOPLE-POWERED CAPITAL

CARMEN ROJAS

As we begin to realize the promises of platform cooperativism, the challenge we face is not a lack of imagination or of ideas for generative enterprises. We also don't lack people who are motivated to create platforms, networks, or products for what we need to become better, more complete, and deeply loving human beings. We don't lack leaders, doers, technologists, or organizers. What we often *think* we lack is capital. But this doesn't need to be the case, and there is no time like the present to prove that the resources exist to make platform cooperativism real.

I speak from experience. I work for The Workers Lab, a labor-backed innovation lab focused on supporting and investing in organizers and entrepreneurs to create enterprises that have the promise to build power for working people. We recognize that the United States has become a low-wage nation where millions of workers invest hours into jobs with little hope of meeting their basic needs. The result has been a social and economic crisis marked most distinctly by the meltdown of 2008. We believe that workers should be able to thrive and play an active role in our economy and democracy. We've even learned some lessons about how to make this happen.

THE STATE OF PLAY

There is a perceived mismatch between ideas and capital in the conversations surrounding platform cooperativism. Many people have great ideas for creating platforms that promise to powerfully connect people with what they need. But they often struggle with testing their ideas or developing a prototype because they don't know where to find the capital necessary to shift from ideation to creation.

I would argue there are two reasons for this perceived mismatch between ideas and capital. The first is that well-meaning technologists and entrepreneurs often are not connected with those who might want to use, adopt, and own their platforms. This leads to the creation of boutique solutions distant from the people who would most benefit and most likely pay for their existence. If there were a closer connection between the creators and the potential users, there would be more ways to resource, create, and validate platforms.

The second reason for the perceived mismatch between capital and ideas is not unrelated to the first: the expectation of somehow obtaining traditional capital resources to support work that ultimately challenges the existing system. Instead, we need to tap into a radical imagination that starts to undo the alignments of power that dominate the financial landscape. To do so, we should begin with ourselves.

WHO IS THE "WE"?

The creators of platforms and their end users are usually very different. Most people today who are exploring the creation of platforms co-ops are still white men who have few to no relationships with those they would like to see governing and owning the platforms they are creating. If they're serious about considering their users as true members and partners, the perceived capital mismatch is actually a *customer* mismatch.

This reminds me of an experience we had at The Workers Lab. A young white entrepreneur had created a platform to connect recently trained truckers with quality jobs with benefits. Still new to the world of investing, we were anxious to prioritize employers' needs over those of workers, based on an unspoken assumption that any job was better than no job. We approached the entrepreneur, and he was easily swayed to refocus his business to meet the demands of employers, as opposed to persisting in his ambition to connect skilled workers to quality jobs that they could validate.

This is not an uncommon story. The promise of platform cooperativism is to create something new, where those who are most marginalized and exploited on traditional platforms have another place to go. But the allure of old-fashioned capital can close that door before it even opens.

A number of leaders have recognized this mismatch and are working tirelessly to address it head-on—for instance, the creation of a training, monitoring, and certification enterprise by the Workers Defense Project in Austin, Texas. After organizing for ten years to address the rampant exploitation and lack of protection among immigrant workers in Texas' construction industry, it created a business where immigrant workers are both the providers and beneficiaries of safety training, can act as on-site monitors, and have co-created the terms to certify construction projects. By investing in communities of color and immigrant communities to act as leaders within their business, they are demonstrating that the "we" of this work can look like the "we" of the world.

WHOSE MONEY?

It is possible that we could end up creating platforms that are owned by the users without truly aligning them with our values. True alignment means aligning the enterprise with the capital that drives it.

At The Workers Lab, we've seen a number of reasons for the difficulties that people have in creating bridges between their ideas and the resources they need. For one thing, those of us who see ourselves working on the side of justice have often become convinced that capital can't be a positive tool in this quest. We've heard this time and time again from the kind of organizer who is meanwhile funded by major foundations, which are subsidized by the state as nonprofits. Despite the contradictions involved, eschewing capital has become a matter of activist identity.

Others, meanwhile, expect that the same sources of capital that back exploitative platforms should or would fund their more cooperative work. This prospect is not only unlikely, it is also destructive. Such an approach reduces our expectations to changing the rules of the game without changing the game itself.

At The Workers Lab, we are looking to marry our ambitions of platform cooperativism with cooperativist finance. The Securities and Exchange Commission has enacted new rules on crowdfunding, through the JOBS Act, that are opening up the ways that everyday people can invest in the companies that best align with their values. These new rules allow non-accredited investors to invest, and they also allow for investments to occur without the intervention of brokers. It has become easier, in this way, for us to be responsive to the people we hope to serve and partner with.

Of course, these rules will also allow for the creation of and investment in more exploitative platforms. But our ability to expose those who are using this opening for those purposes can be addressed by organizing and exposing the bad actors. Doing so will help align us with our collective worries and dreams. When we are the ones investing in us, new kinds of change are possible.

As we look back at the movement for platform cooperativism years from now, my hope is that we will see a moment that stemmed the surge of platforms that exploited working people, devalued human well-being, and placed profits over all else. My hope is that we will have created alternatives that became the norms for how platforms could be used, owned, and shared. My hope is that we will also have, more broadly, reset the terms of capital and creation for our world.

32. CAN CODE SCHOOLS GO COOPERATIVE?

KAREN GREGORY

As public universities continue to weather attacks on funding and on their curriculums, the phrase "learn to code" has become something of a mantra in higher education. It is as though uttering the words to students might be a cure-all for what ails the current job market. Across the United States, public universities are currently struggling to revise and revamp curriculums in the face of the infamous "skills gap"— a gap that is presumed to exist between what college students currently learn and what skills are required in the market. Despite evidence that such gap claims are overblown, the narrative and rhetoric of the skills gap, which banks on a feeling that university curriculums have not kept pace with advances in digital technologies, has nonetheless become a powerful weapon in the assault against the public university. Yet, while much ink has been spilled over the value of learning to code, the conversation seems to miss a fundamental issue: learning to code is often entangled in a larger privatizing, entrepreneurial mission. If we stand any hope for an equitable digital future, we will need to situate a new figure of labor at the heart of learning to code. We will need a new figure of labor working to design and develop digital platforms.

Yet, while public universities struggle to find their place in the emerging higher education economy, a new entity has emerged on the scene—that of the code school, which offers an independent, stand-alone set of intensive programming courses to students. I would like to argue that we can learn a lesson from these code schools, but we can do them one better. We can bring such projects into the public university as a way to teach new forms of cooperativism, as well as contextualize such digital work in the larger story of labor history.

Code schools, such as General Assembly, which is a global provider of intensive digital skills courses, currently offer courses such as a twelve-week "full-time career accelerator" training program in Android development. They do this at an additional cost to an individual student, as code schools currently stand apart from the traditional liberal arts students. The schools explicitly link coding and career. Udacity even goes so far as to guarantee job placement to students in their machine learning courses. If students do not land a job after they complete the course, they are entitled to a refund on the cost of the course. Rather starkly, Udacity's founder Sebastian Thrun has been quoted as saying, "The ultimate objective of education is to find people a job."

Code schools strike a chord and they capitalize on a relatively recent social development—the emergence of the educated underemployed or unemployed. Code schools are able to profit from a relatively well-educated pool of potential students who are now job seekers in a depressed market and who are willing to pay out of pocket for additional skills. In many ways, the existence of the code school legitimately raises the question of what "skills" colleges and universities should be imparting to students. If code is to be our shared lingua franca in an increasingly digital and connected world, then more of us will need to learn to speak in its vernaculars not only in order to "get to work" but to continue the necessary project of critical, theoretical scholarship. Coding or programming skills need not be counterposed to the work of thought, despite the insistence of some in Silicon Valley.

Yet what interests me here is not debating the merits of learning to code. What interests me, rather, is the social figure of the private (and entrepreneurial) laborer who is often brought into being by such coding projects, particularly when such projects are brought into the public university and sold back to administrators as a way to embolden their curricular offerings and recruit students. When the code school returns to the public university, perhaps as a "coding boot camp" or a "startup incubator," it is with the explicit expectation that students will be taught how to be entrepreneurs—how to embrace and identify with an entrepreneurial model of work and with an expectation that private risk and debt will eventually be rewarded with the riches of the entrepreneurial digital market. Yet if the educational context and conditions of learning to code simply reaffirm the privatization of risk and debt, then what expectations can we have that the digital world will reflect more than profit interests? Furthermore, such an emphasis on entrepreneurialism threatens to duplicate the known gendered and racial biases and structural inequalities that already plague the digital economy.

As long as the context of learning to code or program requires an embrace of the heroic figure of the lone, privatized individual, the digital world will continue to be built in its image—it will continue to reify the very notion that the digital economy is a site of "ownership" and "reward" rather than collective, shared, and public resources.

While the language of digital "making" has entered the university, it has come with a curious disinterest in labor and labor history, which, for all practical purposes, have been radically abandoned in the contemporary university. Yet, as the university continues to grasp at entrepreneurial straws, it overlooks an opportunity to reconnect public higher education to the (rather unknown) future of work. While coding may sound like a solution to current market woes, in truth, it will only delay the eventual degradation of such digital labor. When everyone can code, those jobs too will go the way of other forms of work—outsourced, undervalued, underpaid, or automated. However, a digital skills curriculum anchored in labor theory and labor history—a curriculum that explores the possibilities of new forms of collectivities, organizing, and worker agency—has the potential not only to generate a new app or platform, but to reconfigure how digital labor is brought into being and how we imagine continuing to live in and through digital platforms, networks, and infrastructures.

Such a digital skills program could be built in and through working with institutes such as the Murphy Institute at the City University of New York. Students would be introduced to a broader world of labor organizing, cooperative structures, community rights, and labor politics as they also learn to design, develop, and program. It is also essential that such a program be designed in and through the public university as it is precisely public university students, such as CUNY students, who currently stand very little chance of entering into the elite world of Silicon Valley. A cooperative code school would not only help to resituate the public university as a vital tie between students and the failing job market, but it stands to offer a chance to rethink the social demographics of digital labor.

Code schools and their ilk are emerging at this time because they are responding to a need for a trained workforce, but they are succeeding because they tap into the very real sense that the labor market in the United States, even for the highly educated, is broken. As coding projects return to the public university in the guise of entre-preneurialism, we are losing a valuable opportunity to rethink how digital "skills" might become a fertile ground for a new, potentially cooperative-based, labor consciousness. It is time to bring labor history back to the classroom so that students may begin to rethink technology and infrastructure in order to build a more equitable world.

33. A CODE FOR GOOD WORK

PALAK SHAH

Once every few generations we have a momentous opportunity to make big changes, to reset norms and culture and to reinvent ourselves. That's why the emerging platform cooperativism movement—as an opportunity to create an economic system that is empathetic and efficient, a system that solves for equity—is an exciting and radical development. As we embark upon building these new models, defining our core values can provide useful guidance to the emergence of a new economic system.

The essence of platform cooperativism is a rejection of uneven extraction and an emphasis on cooperativism. This movement is rightly asking: Who are we problem-solving for? What is the impact of the new economy on the workers who are the engine that drives it? Will the new economy be more or less equitable than the old economy? Will there be more empathy in the new economy? Will the new economy work for all of us, or is it an online version of the economy that worked for just a few?

At the National Domestic Workers Alliance, we already know what it looks like when the economy optimizes for some of us and extracts value from the least visible of us. We've been working on promoting respect and dignity for some of our most invisible and under-valued workers in the offline economy for years, and what we've heard from these workers about the online economy echoes those injustices.

Workers in the on-demand economy tell us that they like the flexibility of their work, but they fear being cut off from their platforms without warning after they have to cancel a shift to care for a sick child. Many workers like the economic possibilities of these emerging sources of revenue, but instead found themselves being paid less than minimum wage to clean a completely trashed house.

So far, we have little evidence that Silicon Valley will build an online economy optimized for equity and dignity. While new platform technology presents us with entirely novel ways of organizing and working together, and a strategic opportunity to reimagine our economic relationships to one another, the tech businesses have largely focused on enriching investors' returns and delighting customers, often at the expense of the workers who power these platforms. Algorithms can handily resolve supply and demand imbalances or competing customer preferences, but the ruthless pressure to pursue growth and profit still overpowers the moral case of treating workers well, limiting how equitable these models can be. While our technological advances can make life easier for all humans—isn't that the point of technology?—it is not yet clear that we will use it to design an economic system that works for all of us.

What would it mean to build platforms that go further than efficiency, style, or convenience? How would we integrate compassion, dignity, or fairness? How would we write algorithms that optimize how we treat one another, how we truly connect, and how we sustain the very people that sustain us? How do we write that into operational code?

It's important that we ask these questions, but it's vital that we answer them. A political agreement on collective ownership or ideological agreements on worker power can be the first, essential step. But the real work is in building models that pick up where commercial platforms have left off. That means we must roll up our sleeves and dive into the operational details and inner workings of how labor is distributed through algorithms, phone-based apps, and online platforms.

This is not as easy as it seems, and there are no shortcuts. The central task here is to build platforms that can operationalize our commitment to fairness and cooperativism. Why? Because our values alone won't get us all the way there. Building a cooperative platform is also a pragmatic, operational endeavor, and the devil is in the details of seemingly unglamorous and mundane processes.

In many areas of business, technology, and even manufacturing, organizations rely on certain agreed-upon standards to establish a set of common operating assumptions describing *what* the system should do—not necessarily *how* the system should do it. Whether you're talking about platform co-ops or Silicon Valley giants, we believe a set of values-based specs for the new economy could serve as a guide to embedding dignity and respect into all operations areas—a Good Work Code for the new economy.

Our Good Work Code is a set of eight simple principles that can serve as a framework, a guide ensuring that the new platforms are creating good work:

Safety: Good work does not allow for us to wonder at the start of a shift if we will be unharmed by the end. Everyone deserves to be safe at work, always.

Stability & Flexibility: Good work is made possible when we are not anxious about meeting life demands, whether it's making an unexpected doctor's appointment or making enough money to pay the bills. We are all at our best when our schedules allow us to balance work and life with a stable—but flexible—schedule.

Transparency: Good work means being transparent about requirements, performance, and the rules. When everyone knows how things work, everything works better.

Shared Prosperity: Good work rewards all of us. Workers are the engine powering the platform, and when the platform thrives, they should thrive too.

A Livable Wage: Good work provides a living. That's why we work—to live. Everyone needs fair pay and benefits to make a living.

Inclusion & Input: Good work recognizes that our value extends beyond performing a task. Platforms are more successful when we are heard, respected, and valued as part of the team.

Support & Connection: We all work better when we don't feel isolated and alone. Good work supports us to adapt and manage in a rapidly changing environment and economy.

Growth & Development: Good work provides opportunity for the most fundamental human need: to grow. Everyone wants to grow and learn at work as they do in other areas of their life.

The Good Work Code doesn't introduce any revolutionary new values. It doesn't innovate or disrupt previous thinking on good work standards. But this framework, or a derivative of it, will be essential to building an equitable platform. We are at the exact right moment in time to insert this framework into the DNA of the new economy, to correct the course so that we're building a new economy that works for all of us.

With the emergence of the platform economy, we have one of those rare opportunities to reset the norms and culture of work. At its best, technology strengthens our humanity and our connectedness. We shape our economy as much as it shapes us, and we have an opportunity to make sure the digital revolution is supported by a long-overdue revolution in values.

34. MEET YOUR FRIENDLY NEIGHBORHOOD TECH CO-OP

MICKY METTS

When I joined a web-development tech co-op, it changed my life in wonderful ways. I am now able to bring my whole self to my work and have a life that is not divided between work and play. My drive and passion come from doing what I love—building community, and enabling people to have the technical tools they need to maintain and control their autonomy.

It is rapidly becoming easier for someone with a great idea to build a company online without much of a barrier to entry. Websites can be set up for free or minimal cost, and cloud services with online website building tools are plentiful. But sophisticated, cutting-edge platforms still require skilled people to build them. A new generation of platform co-ops will need developers who understand both technology and cooperative enterprise. Nobody is better prepared for this work than tech co-ops.

What, you might ask, is the difference between tech co-ops and platform co-ops? How might they collaborate?

Tech co-ops are worker-owned development shops that build customized tools. Platform co-ops are online tools owned by the people who use them. They are different, but they will both be stronger if they work together.

As a co-op and as individuals, Agaric is part of the vibrant, global Drupal community, which maintains and continually improves a free content management system that makes it easier for people to build websites. Some of Agaric's projects result in new software that can be helpful to others, or solutions to challenges that others might be facing. When this happens, we package the software we create for free distribution, usually on Drupal.org. Supporting people in building websites that make the world a better place is rewarding, but being able to package that free software and put it in the hands of others is nothing short of incredible. We also collaborate with groups such as the M.I.T. Center for Civic Media to brainstorm and build projects that benefit our neighborhoods.

Agaric is made up of five people. We are spread around the world, but we are all working on a level playing field. Being a cooperative allows us to govern ourselves through bylaws that we created. Worker-owners can join by paying a small, monthly investment, of which a portion can be recovered if they decide to leave. We earn the same wage each month, and we all have overlapping skills that complement each other. We take time to teach each other what we learn and broaden our knowledge through tutorials and collaborations with other developers. Freelancers can gain a lot by putting their efforts together and building small cooperative companies that can provide services to their communities and beyond.

Like other cooperatives, tech co-ops cooperate with each other. I am part of the Tech Co-op Network, which lists more than twenty-five tech co-ops as members on our website, TechWorker.coop. We have a mailing list where members can post messages. We share information about our projects, and ask each other for advice or links to resources such as example bylaws. The group is a good source for recommendations on governance and details of how successful cooperatives are structured. The list is also populated with people who are not yet part of co-ops, but would like to form or join one. It provides an informal introduction to cooperative business and makes it easy to ask questions and stay informed.

Being part of a cooperative allows the members to have a voice and bring projects to the group. When one of our own presented a sound proposal, we started Agaric Nicaragua, a project to build a resource for developers in Managua to learn programming skills and to be trained to take on development roles with clients. This is a win for developers in Nicaragua and for Agaric, to grow talent in areas that are underserved by local educational opportunities around the world. We will extend this opportunity to the world by sharing the template for this project.

This kind of collaboration is essential. Creating a cooperative internet will require more than just new technology. It will no doubt take the combined efforts of many tech co-ops to build a new ecosystem of cooperative platforms. The structures and processes we use to work together must also change from hierarchical, linear forms to non-linear, cooperative ones. To give a sense of what that entails, here are some steps for starting a tech co-op of your own:

1. **Find the right people.** You will need to find coworkers in your industry that value working on a one-worker-one-vote basis. Talk to people in your personal network about your goal. Let former coworkers know you are forming or seeking to work in a cooperative. Reach out to mailing lists you are on and ask if people are interested in working collectively.

2. **Explore different strategies for self-management.** Learn about your local cooperative community. If possible, go to events where you can meet members of cooperatives (tech or otherwise) and ask them how their organizations are structured internally. Most cooperative members are approachable and willing to answer your questions. Ask them what they think works and what doesn't.

3. **Consider conversion.** This can be easier than starting from scratch. If you work for a company you like that is not a cooperative, talk to the owners about the possibility of selling it to the workers.

4. **Define the parameters of your cooperative environment.** Whether through your articles of organization, bylaws, or a simple

contractual agreement, be clear about how your cooperative will work. Above all, a cooperative is defined by its members.

5. **Join a cooperative network (or two).** For those in North America, consider the U.S. Federation of Worker Cooperatives and the Tech Co-op Network. Wherever you are, connect with whatever kinds of co-ops exist in your area. If there isn't already a directory of local co-ops, consider starting one!

6. **Invest in other cooperatives.** Buy locally whenever you can. Encourage pooled funds from successful cooperatives to help bootstrap new proposed cooperatives. When you invest in new local cooperatives you are investing in your community. Agaric often collaborates with other co-ops, and we share our leads with a pool of developers that we have worked on projects with in the past. Once you have a healthy network, word-of-mouth referrals go both ways.

7. **Choose free tools to run the business.** Free software is software that can be used, studied, modified, and redistributed by anyone, for any purpose. Using free software to run your cooperative is not only a way to preserve your freedom, but it will allow you to share your successes with others by sharing your code. We recommend free software tools like GnuCash for accounting because it doesn't require you to trust a corporate cloud service. If you think of ways to improve it, also, you can get involved with the actual development if you choose to make changes. At Agaric, we believe that cooperativism and free software go hand in hand.

Through steps like this, we really can make our own jobs, manage our own time, and create our own online platforms through cooperation. Co-ops can teach communities how to be freer from oppressive systems by being role models. But we must be wise in how we do it. Strong tech co-ops will make for stronger cooperative platforms. Co-ops must work together in solidarity.

35. BUILDING THE PEOPLE'S OWNERSHIP ECONOMY THROUGH UNION CO-OPS

MICHAEL PECK

Cooperatives and unions started out their organized lives together. In gritty Northwest England, during the Industrial Age's heyday—Manchester in the 1840s—the Rochdale Society of Equitable Pioneers composed their Rochdale Principles, which became the foundation of the modern cooperative movement. It was in this same region, around the same time, that industrial labor unions were beginning to flourish. Since then, diverging histories, experiences, and destinies have caused cooperatives and unions to run along mostly autarchic paths, sometimes parallel, too infrequently connecting. Now, as we find ourselves in a period of the highest landlord absenteeism since America's Gilded Age, it's time to bring unionism and cooperativism back together.

America's inequality epidemic calls out for a new labor-cooperative convergence to accelerate the end of the era of false structural choices. Yesterday's Hobbesian, "either-or" menu, forcing workers to choose between sustaining jobs or a clean environment, or between racial justice and secure employment, is being replaced with a more positive and uplifting "all of the above"—often through online platforms facilitating better choices and stronger, more inclusive advocacy. Compartmentalized hierarchies are breaking down into a more egalitarian digital commons, overcoming imbalances between labor and capital. Emerging communities of freelancers are exhibiting what Sara Horowitz, founder of the Freelancers Union, calls the "New Mutualism."

Increasingly, these new pioneers prefer to own their labor and rent their capital whenever possible, instead of Wall Street's predatory opposite. With less than 10 percent of Americans currently owning their own businesses and workplaces, today's "new, new organizing" begins to address the skewed imbalances between capital and labor and the power this distortion produces and exercises. Movements and policies are seeking to extend the "sharing economy" into an ownership-enabler, and to resist manipulative downgrades into a rental economy, where labor becomes even more of a bottom-traded commodity without equity benefits.

Yesterday's Davos-sanctioned global marketplace for labor arbitrage is becoming unmasked as a cabal for corporatist buying and selling of human beings and a contributor to global inequality, with the excuse that labor is a disposable commodity instead of a precious resource. This turnaround is long overdue.

Starting in the spring of 2014, 1worker1vote.org set out to demonstrate that widespread workplace equity and democratic participation can return America to its original system of individual and local community ownership. This initiative emerged from the historic 2009 United Steelworkers collaboration with Mondragon in the Basque region of Spain, the world's largest network of worker cooperatives. We're now developing a nationwide cadre of unionized, worker-owned-and-managed cooperatives to overcome domestic structural (racial, gender, geographic) inequalities of opportunity, mobility, and income. A growing multitude of like-minded local and national organizations are working with us to help existing businesses transition to a union co-op structure and to launch new union co-op businesses.

Our threefold intent is to:

1. Defeat embedded structural inequalities by deploying tested and proven hybrid ownership models (starting with the union/co-op template);
2. Build and launch profitable, worker-owned-and-managed enterprises in the context of inter-cooperating ecosystems, drawing on sixty years of the Mondragon experience;

3. Co-design projects and tools that can replicate and scale (the complete 1worker1vote.org mantra is: "include, design, build, launch, replicate, and scale").

As a clear case in point, and after almost eight years of struggle and action, Abdi Buni, the president of Denver's Green Taxi Cooperative, and Lisa Bolton, now the Communications Workers of America's international vice president for telecommunications and technologies (but previously president of CWA Local 7777 in Denver), have earned the right to claim success. As a result of their determined "learning by doing" process and teamwork, unionized and employee-owned cab companies will dominate the taxi marketplace in Denver.

This represents a remarkable example of how solidarity-centric business structures can combine with determined union policy advocacy and market-available technology to produce more holistic business models. Green Taxi Cooperative has more than eight hundred worker-owners, who are also CWA members, with a newly signed collective bargaining agreement. Before this, CWA Local 7777 had assisted another local taxi cooperative, the two-hundred-driver Union Taxi, which is presently not unionized, and the union learned a lot from that process during round two.

Green Taxi is joining other taxi drivers in New York City and elsewhere who are launching apps to step up the competition with Uber, allowing customers to hail a cab and pay for it with their devices of choice. The platform cooperativism movement is poised to combine local solidarity structures with open-source technology, transforming the Green Taxi precedent into a repeatable and scalable economic opportunity nationwide.

These opportunities are even greater now that the California Labor Commission has ruled that Uber drivers are employees, not independent contractors. In more and more jurisdictions, app-based drivers are winning the right to form a union. This makes it even easier to align platform cooperativism with the might of organized labor in the transportation sector.

Other emerging economic networks are beginning to aggregate as well. The Freelancers Union, headquartered in Brooklyn, is the nation's largest labor organization representing the new, independent workforce of 260,000 members in more than twenty U.S. cities—as well as the 53 million Americans freelancing today, more than one in every three workers. In four years, the American Sustainable Business Council, based in Washington, DC, has become the nation's fastest growing private sector trade association of over 250,000 businesses and 325,000 business leaders focused on building a sustainable economy based on triple-bottom-line principles: people, planet, and profit. These inter-cooperating communities are forging private and non-profit solutions in which enlightened government is a minority partner but not a market-maker.

We can start to see firsthand how new hybrid models such as union-cooperatives can place worker ownership in the economy's center ring, enabled by technology. The online "commons" can reinvigorate the American Town Square and Main Street, benefiting from strong antecedents. Historically, marginalized communities such as African Americans in the Deep South's agricultural counties were cooperative pioneers because, as in the churches, they had their own spaces to democratically organize. Meanwhile, people in racially and economically conflictive zones such as Ferguson, Baltimore, and Detroit are rediscovering that to secure their civil rights, they must also secure local equity and ownership.

Research demonstrates that positive employee and company performance over time correlates with high-impact participation on all levels by workers, combined with the broadest possible equity distribution among workers and a strong emphasis on worker education. Employees with some form of worker ownership accumulate more savings than employees in non-participating firms. Firms with some form of capital-sharing perform better in the competitive marketplace than those without. Workers with profit-sharing or employee stock-ownership and stock-purchase plans are better paid and have more benefits than other workers. These kinds of firms also weather economic downturns better than their investor-controlled counterparts.

Despite this evidence, the so-called American Dream of widespread ownership is receding for a rapidly diminishing and resentful middle class. But grass, as Pete Seeger sang, still grows through concrete. Enabling workers to become owners in their projects and companies reflects core American values of freedom, individual dignity, self-reliance, bootstrapping, solidarity, and equal opportunity—all reinforced by productive ownership principles and practices. These ineluctable, hope-instilling values promise an economy that gushes up rather than trickles down, an economy that is not rented, tithed, leased, outsourced, or off-shored—an economy in which every participant, every worker, has a voice, a vote, and a right to share and participate in the common good.

36. TOWARD A THEORY OF VALUE FOR PLATFORM COOPERATIVES

MAYO FUSTER MORELL

Collaboration through online platforms does not happen independently of their design and governance. Whether or not they are thriving depends on platform ownership. Such ownership might condition how well a particular platform can trigger participation and community interactions, and yet it has rarely been investigated. When was the last time you heard someone ask about platform ownership?

The Platform Cooperativism event at The New School in 2015 did focus on ownership, but even there quantitative analysis of platform data was largely missing. With regard to user engagement and value creation, does cooperative ownership really make such a big difference to corporate ownership?

This chapter draws on research from the P2Pvalue.eu project, which investigates the conditions that favor value creation in the context of commons-based peer production. Our analysis, which examines three hundred European organizations, cannot claim to be representative, due the fact that we cannot account for the entirety of the very diverse universe of commons production. We did, however, at least try to represent the heterogeneity of the field—from free, libre, open source software (FLOSS), and open data, to open design and open hardware. Our results are freely available at directory.p2pvalue.eu.

Commons-based peer production refers to a set of activities characterized by collaborative production, involving peer-to-peer relationships and a resulting common resource. This model stands in stark contrast to traditional, hierarchical command relationships. It relies on access to open commons resources—favoring access, reproducibility, and emulation. Some of the best-known examples are Linux, Wikipedia, OpenStreetMap, and SETI. Commons-based peer production is not identical with platform cooperativism, but our study is relevant to the larger platform co-op ecosystem, which is deeply reliant on commons-related practices.

For a community of commons-based peer producers to operate, there needs to be a platform—made possible by the people who create it, maintain it, and facilitate its legal framework. Notably, there are different types of platform providers. Wikipedia, for instance, is facilitated by the Wikimedia Foundation, a nonprofit foundation steered by the Wikipedia community. In other cases, however, the platform provider is a corporation, which leads to far less conducive conditions for the community. One example is Flickr, the photo- and video-hosting site owned by Yahoo.

In our research, we identified four types of platform providers: public institutions; corporations; nonprofit organizations such as foundations, associations, and cooperatives; and informal grassroots organizations that may not have any legal status. Among the communities we studied, 7.2 percent relied on public institutions, 29 percent relied on corporations, 57 percent relied on cooperatives, and 6.8 percent chose grassroots organizations or community networks as their platform provider.

GOVERNANCE OF COMMON PRODUCTION

Studies of commons-based peer production have usually investigated specific, isolated features linked to the governance of such production. In contrast, we adopt a holistic perspective to understanding the control and direction of a platform, as well as the distribution of power. We considered the following six interrelated factors as determinants and drivers of commons governance:

Mission. It really matters who defines the mission of a platform. Depending on whether it is members of the community or corporate owners, the platform will develop in accordance with the initial mission statement.

Management of contributions. This refers to the extent to which participants are able to decide their own level of commitment, whether they can define their contributions based on their personal interests, motivations, resources, and abilities, and whether relationships are peer-to-peer in contrast to traditional forms of hierarchical command. Greater flexibility of participants seems to be conducive to higher degrees of contribution.

Decision-making with regard to community interaction. Governance of commons production depends on decision-making bodies that also address conflict resolution. Consensus-based decision making is frequent in commons-based peer production but the methods differ.

Formal policies applied to community interaction. As it evolves, commons-based production tends to establish formal rules that may be restricting. Such rules include the terms of use and intellectual-property licenses.

Design of the platform. Individuals are rarely involved in direct dialogue or negotiations among themselves. Instead, they interact with the platform design, which steers their participation and interaction. Therefore, the design must follow the social norms of commons-based peer production.

Platform provision. We noticed two main axes of platform provision: open versus closed with regard to community involvement, and user autonomy versus dependency. A platform is considered open when participation in the provider space is possible for anyone. Participation in these cases is regulated through self-selection. For participants, autonomy is linked to the license held for the commons-pool resources and the type of software used for the platform (i.e., copyleft licenses and the use of freely available code, as in FLOSS, versus conditions defined by ordinary copyright). If the platform can be replicated—if it is "forkable"—the relationships created on forked versions are free from the original platform provider. FLOSS and copyleft licensing allow platforms to be replicated, while close copyright license regimes prohibit that. In other words, the use of FLOSS and a copyleft license creates conditions in which the community can have greater autonomy and freedom from the platform provider.

Governance very much depends on who is in control of these six power nodes in commons production. Each of the axes of governance can be managed in an inclusive or exclusive way. They may encourage involvement on the basis of participants as individuals or through the community as a whole.

The emerging varieties of governance for commons-based peer production are highly complex. We found that specifically cooperative ownership leads to more self-governance among all involved and a more horizontal relational structure; it favors a more peer-to-peer oriented process.

VALUE PRODUCTION FOR COOPERATIVE MODELS OF PRODUCTION

Studies of commons-based peer production have not produced a consolidated analytical framework to assess the value that is produced. One fundamental challenge for the development of such a theory of value is the inadequacy of monetary metrics as proxies for value production, since part of what peer producers create, exchange, and consume does not pass through monetary exchanges. Thus, as part of our project, we are developing our own conceptual framework that identifies six dimensions to assess and measure value:

1. Community building
2. Social use-value of the resource created
3. Reputation
4. Achievement of the stated mission
5. Monetary value
6. Ecological value and derivative processes

GOVERNANCE AND VALUE

Does the type of platform provision—that is, platform cooperativism versus platform capitalism—affect the collaborative community's capacity to generate value? We found that cooperative structures have a positive impact on value creation in terms of use-value and reputation. The cooperatives ranked better on web-based use value and reputation value indicators (such as Google PageRank and the global Alexa ranking), and performed less well on reputation value indicators linked to social networks reputation (such as Kred, which is linked to Twitter, and likes on Facebook). Nevertheless, our analysis didn't find any significant correlation on monetary value, capacity to achieve its mission, or capacity to build community.

This is a preliminary analysis that we will elaborate further as part of the ongoing P2Pvalue.eu research project, but a few things are clear. What we have shown so far is that platform ownership models do not only determine governance but also influence the capacity to generate value. This should be considered when peers or policy makers decide what kinds of platforms they are going to build or promote. According to our analysis, platform cooperativism will further not only self-governing processes but also the capacity for people to create resources and services that garner a good online reputation.

37. PUBLIC POLICIES FOR DIGITAL SOVEREIGNTY

FRANCESCA BRIA

The scale of the transition to platform capitalism is massive. The builders of emerging online platforms aim to become pervasive across all productive sectors, and to permeate every level of society: the level of the individual (with smartphones and wearable technology, lenses, glasses); the level of the home ("smart homes," smart power meters and Internet-connected sensors); and the level of "smart cities" (driverless cars, networked transportation services; energy grids, drones, ubiquitous digital services). Platforms are reshaping not just the Internet but the economy as a whole, and governments have a responsibility to ensure that this new economy serves more than the platform-builders' profits.

We are seeing a shift of power, for instance, from service intermediaries to information intermediaries, a kind of "Uberization" of services. The current data-driven platforms are marketplaces that match potential customers to anything and anyone. They are able to gather lots of data, lower transaction and coordination costs, and provide cheaper services using a dynamic pricing strategy. Most platforms are monopolies, quickly capturing network externalities by exploiting the network effect and the economies of scale of their ecosystem. They are also parasitic, since they free-ride on collective data and people's existing social and economic relations. The strategy of these powerful algorithmic institutions is to enter a variety of economic sectors rapidly and disrupt current industries. By controlling their digital ecosystems, they can turn everything into a productive asset, and every transaction can become an auction where they set the bidding and pricing rules.

Platforms are also increasingly transforming the labor market. Uber, for instance, does not own cars and doesn't employ drivers; it regards its workers as independent contractors. In this way, the company externalizes most costs to workers, eliminating collective bargaining and implementing intrusive data-driven mechanisms of reputation and rankings to reduce transaction costs (for the company). The growth of the sharing economy has so far come with an increasing precarization of labor, and erosion of job security, social protection, and safety nets for workers, such as benefits related to healthcare, pensions, parenting, and so on.

If you are European like myself, and you're used to a functioning, social-democratic safety net, what is now promised by companies like Uber and Airbnb is not very appealing. Despite their optimistic pitch of delivering better and cheaper solutions to solve the world's greatest problems—from climate change to health and education—the welfare program offered by Silicon Valley comes with public services cuts, austerity policy, financialization of public infrastructure, increasing debt, and a free license for tech corporations to monitor citizens twenty-four hours a day.

Many would argue that the European welfare model is no longer suitable or sustainable. However, there are historical and political reasons that got us to the current situation. Governments forced to implement counterproductive austerity measures are left with no budget to invest into social policies.

A common rationale used in defense of the platform economy is that it will generate a huge wealth for the platform owners, and they will reinvest these profits into the real economy, thus serving the public good. Unfortunately, this is not the case. On the contrary, the latest wave of digital innovation has resulted in excessive returns to capital, with massive amounts of cash going to the balance sheets and the offshore accounts of big tech companies, while very little gets invested in welfare, social infrastructures, education, health, and clean energy to fight climate change. This situation is exacerbated by the apparent inability of governments to tax profits made by high-tech and financial giants, as seen recently in the very generous tax settlement between Google and the UK government and the tax dispute between the Italian government and Apple.

THE SHIFT TOWARD DEMOCRATIC, COMMONS-BASED CITIES

The search for alternatives to platform capitalism should be put within a broader framework of growing discontent with austerity measures and the corporatization of everything. In Europe we have very good examples of movements advocating for the collective management of public resources such as water, air, and electricity. These represent potential alliances when we discuss cooperative platforms.

A very interesting example of a city that is putting forward alternative policies and forward-looking regulations is Barcelona. After the large mobilization of the 15M movement beginning in 2011, the anti-eviction housing activist Ada Colau, a leader in the Platform for People Affected by Mortgages (PAH), became the mayor of Barcelona, representing the main political opposition against the elite who brought Spain into a deep financial and social crisis, which left hundreds of thousands of families without a home.

The new coalition led by Colau has been crowdfunded and organized through an online collaborative platform that aggregates policy input from thousands of citizens. Soon after taking office, the coalition members embarked on a series of radical social reforms. In particular, they started to enforce regulations to block illegal tourism. The council froze new licenses for hotels and other tourist accommodations, promising to fine firms like Airbnb and Booking.com if they marketed apartments without being on the local tourism register. Barcelona then provided these companies the possibility to negotiate 80 percent of the penalty if they allow the Social Emergency Housing Consortium to allocate empty apartments to residents with subsidized rent for three years.

The city has called for a popular assembly for responsible tourism where citizens can discuss best practices and business models. The new government is also promoting new policies to foster a collaborative economy that generates social benefits locally. Besides these types of initiatives, Ada Colau has also promised a shift toward re-municipalization of infrastructure and public services. This is grounded in a very critical understanding of the neoliberal, surveillance-driven "smart city" model being promoted by big tech corporations. The ambition, instead, is for a shift to a democratic, green, and commons-based digital city built from bottom up.

This vision of re-municipalization of critical public services and network infrastructures is of growing global appeal, leading to a new alliance between public utilities and cooperative online platforms. A number of cities and regions across the world are attempting to put water supply, waste disposal, and energy provision contracts back into public hands, prioritizing community interests over private commercial objectives. An important innovation has been the growth of new forms of public utility ownership, combined with more decentralized forms of collective ownership—including cooperatives with shares held jointly by the local authorities, labor unions, and citizens.

COOPERATIVE DATA PLATFORMS AND ALTERNATIVE MODELS FOR DEMOCRATIC INNOVATION

We need public investment in future data-intensive infrastructure and welfare systems for the common good. Cities and governments have not yet fully grasped that power lies, today, at the level of data. Only recently have we begun to view online platforms as meta-utilities, with the information layer feeding all other services, rapidly changing the way services are managed and delivered. Data, identity, and reputation are critical in the platform economy. Silicon Valley aspires to turn data into a new asset class—a commodity to be sold and traded in financial markets, with property regimes surrounding it. Shoshana Zuboff of Harvard Business School calls this new reality "surveillance capitalism." We have to move from surveillance capitalism to a system that is able to socialize data—such as with new forms of cooperativism and democratic social innovation.

Cities, for instance, should be able to run distributed common data infrastructure on their own, with systems that ensure the security, privacy, and sovereignty of citizens' data. Cities can then invite local companies, cooperatives, civil society organizations, and tech entrepreneurs to come in and offer innovative services on top of that infrastructure. One example is the European Commission's CAPS program, which has invested around €60 million on collaborative and open platforms to pilot bottom-up, citizen-led projects with strong social impact such as the D-CENT project (http://dcentproject.eu), developing distributed and privacy-aware tools for direct democracy and cryptocurrencies for economic empowerment. Initiatives like these can help ensure that the data produced by platforms, devices, sensors, and software doesn't get locked down in corporate silos, but becomes available for the public good. Investing public resources for piloting innovative, cooperative platforms is necessary to enable credible alternatives to the current data paradigms exploited by the dominant platforms—integrating economy, technology, society, and policy, which would otherwise remain fragmented and lead to market concentration and regulatory breakdowns.

The current predatory paradigm is not the only solution. We can harness the technology-driven transformation now under way to improve our society and welfare for the collective benefit. Building alternative forms of public and common ownership for data–intensive platforms will help to create an economy that transcends the logic of short-termism and rent extraction. We are not going to be able to improve welfare, health, youth employment, education, and the environment by leaving the market to do it on its own. We should look beyond immediate commercial gain in favor of long-term value creation for society. Twenty-first century democracy depends on this task.

38. LEGAL AND GOVERNANCE STRUCTURES BUILT TO SHARE

MIRIAM A. CHERRY

To date, the dominant economic narrative for the gig economy has been one in which platform owners extract a share of the income generated from the workers who use their platforms. This is troubling, since many forms of crowd-work are situated at the crossroads of precarious work, automatic management, deskilling, and low wages. Recent lawsuits by workers in the gig economy claiming employee status contain the demand for better pay, hours, benefits, and working conditions. However, these misclassification lawsuits do not seek to change the ways in which the underlying business relationship between workers and platforms are structured.

Platform cooperatives, however, subvert the dominant economic narrative. If workers themselves owned the platforms, then workers would have control over important matters such as wages and benefits. Cooperatives could clear a path toward efficient and convenient use of technology for consumers that simultaneously incorporated fair labor standards. For example, taxi drivers in several cities are working on setting up their own driver-owned platform to compete with the popular Uber app. I want to put this new move toward platform cooperativism into context with the underlying legal structures and also to discuss briefly the challenges to governance that platforms cooperatives will face.

Worker-owned businesses have long existed in the United States, although they have been relatively rare and an exception to the default of the traditional for-profit shareholder primacy model. Many advocates who seek to better the status of so-called shadow ("under the table") workers have long advocated for worker-owned businesses through groups such as worker centers. Why would becoming owners make sense as opposed to unionizing and acting collectively to bargain with an employer? With certain endeavors such as home cleaning, day labor, and home health, there are individual contracts but no one common employer with whom the workers can bargain collectively. Likewise, in the gig economy there are many individual customers using the platforms. As workers continue to struggle in the gig economy, platform cooperatives have emerged as an appealing possible alternative.

On a practical level, what legal tools are available to help those who are trying to set up platform cooperatives? Some states have enabling statutes that set out tailor-made rules for worker cooperatives. However, there is no uniform law across the states, and some states have passed enabling legislation only for consumer cooperatives. California faced this issue and, in 2015, amended its legislation to make it clear that both consumers and workers could form cooperative businesses. That said, even in the absence of a worker cooperative statute, there are other business entities that could provide the appropriate organizational structure for worker-owned businesses. One good choice of business entity for a platform cooperative might be the limited liability company, which combines limited liability with favorable partnership taxation. LLCs may be centralized and run by a group of managers (similar to a board of directors in a traditional corporation) or run in a decentralized way with equal voting, much like our traditional notion of a general partnership. If the operating agreement is properly structured so that the workers are made the members of the LLCs and given management rights, then that should accommodate a worker-owned business model.

Over ten years ago, in a paper appearing in the *UC Davis Law Review*, I noted that business planning techniques (which those who have access to financial and accounting resources routinely employ) could be used to improve the situation of low-wage immigrant women workers. Due to language barriers, immigrant workers often are at the mercy of the managers who arrange the work. In this scenario, immigrant workers often work for depressed wages, are paid under the table, and do not receive benefits. In contrast, LLC structures allowed these same shadow workers to organize and own their own businesses, hiring an English speaker—at a set wage—to work for them, scheduling and arranging jobs. Within an LLC structure, the workers are able to decide what benefits would best serve their members. In addition, as worker-owners who are actively engaged in managing the business and paying taxes, LLC members may have an easier time regularizing the workers' immigration status or, at the very least, not creating a tax liability issue for the workers with the Internal Revenue Service. Finally, the experience of receiving training, and becoming knowledgeable in running a business, can assist workers in taking what otherwise could be seen as a "dead end" low-skilled job and transforming it into a much better opportunity for advancement. Many of the advantages for low-wage immigrant workers inherent in a worker-owned business form could also improve the lot of gig-economy workers.

Another intriguing and potentially fruitful possibility for organizing platform cooperatives would be for the platform to incorporate and obtain certification as a B Corporation. B Corporations are a class of for-profit entities that simultaneously strive to create benefits to the environment, workers, or communities. As such, they operate as a hybrid, straddling the category of for-profit and nonprofit. B Corporations strive for transparency, and investors in such firms understand that there may be tradeoffs—opportunities for profit that may in fact be passed by in pursuit of social-benefit goals. The B Corporation incentives would harmonize well with worker co-ops that already have workers' issues at the very core of their organization and mission. They would also resist the type of "mission drift" of cooperatives that lose their social vision, such as electric co-ops that continue to use polluting coal. To date, eleven states have passed enabling legislation to recognize B Corporation status, with additional states passing similar or complementary types of legislation, such as California's flexible purpose corporation. These business forms put social benefit at the heart of the organization's mission.

Regardless of the choice of business entity, another important issue is designing a workable governance structure in the operating agreement or corporate documents. There are some issues unique to online platform cooperatives that could present particular challenges to governance. Some of the issues include accommodating for flexibility and part-time work. One of the main attractions of the gig economy is flexibility. Worker-owners in platform cooperatives may be working part-time, and there will be a need for ease of entry or exit. Another issue could arise around the amount of effort workers contribute. Although one hopes that workers who work for themselves and other workers will dedicate themselves to building their platform, cooperative endeavors could create moral hazard and the risk of shirking. The other challenge with crowd-work, where the work can be performed in any geographical location, is that there will be participants from many different countries, each with its own set of legal rules.

The fact that there are no tailor-made enabling statutes geared specifically toward platform cooperatives contributes to increased setup costs and barriers to entry. But many businesses that do not fit the traditional mold have had to confront this issue before. Platform cooperatives will be eligible to seek out financial and technical assistance from the same worker centers and legal services agencies that have helped set up worker-owned businesses in the past. Others, perhaps those that seek B Corporation status, may benefit from seeking pro bono legal assistance or accounting advice from for-profit firms that are looking to give back to the community. The basic legal structures for platform cooperatives, while not "off the rack," do exist. They just require the tailoring that legal and financial professionals can provide.

Given the turnover and flexibility of online platform work, the operating documents should be written to allow for relative ease of entry and exit as a member. In addition, the organizing documents must also set up the relationship in a way that sets out what the expectations are for the members, clearly and succinctly. The documents need to include provisions for reducing the share of profits if an individual member is shirking, and also contain clear provisions defining under what circumstances a member or shareholder may be disassociated. In terms of the global or international scope of many platforms, the operating agreement and other documents can be written to provide for choice of law and choice of jurisdiction. Current statutes allow for electronic or remote voting for boards of directors or members, so long as such procedures are set out in the corporate charter or operating agreement. Note that running a business is riskier for the individual worker in a platform cooperative—like any business, the LLC members or B Corporation shareholders run the risk that there will be no profits.

Perhaps the answer to the misclassification lawsuits and the struggle over employee status is to work around it, regardless of the outcome. While not the perfect solution, already-existing legal structures can be modified to accommodate platform cooperatives.

39. BLOCKCHAINS AND THEIR PITFALLS

RACHEL O'DWYER

A blockchain is essentially a distributed database. The technology first appeared in 2009 as the basis of the Bitcoin digital currency system, but it has potential for doing much, much more—including aiding in the development of platform cooperatives.

Traditionally, institutions use centralized databases. For example, when you transfer money using a bank account your bank updates its ledger to credit and debit accounts accordingly. In this example, there is one central database and the bank is a trusted intermediary who manages it. With a blockchain, this record is shared among all participants in the network. To send bitcoin, for example, an owner publicly broadcasts a transaction to all participants in the network. Participants collectively verify that the transaction indeed took place and update the database accordingly. This record is public, shared by all, and it cannot be amended.

This distributed database can be used for applications other than monetary transactions. With the rise of what some are calling "blockchain 2.0," the accounting technology underpinning Bitcoin is now taking on non-monetary applications as diverse as electronic voting, file tracking, property title management, and the organization of worker cooperatives. Very quickly, it seems, distributed ledger technologies have made their way into any project broadly related to social or political transformation for the left—"put a blockchain on it!"—until its mention, sooner or later, looks like the basis for a dangerous drinking game. On the other side of things, poking fun at blockchain evangelism is now a nerdy pastime, more enjoyable even than ridiculing handlebar moustaches and fixie bicycles.

So let me show my hand. I'm interested in the blockchain (or blockchain-based technologies) as one tool that, in a very pragmatic way, could assist with cooperative activities—helping us to share resources, to arbitrate, adjudicate, disambiguate, and make collective decisions. Some fledgling examples are La'Zooz, an alternative ride-sharing app; Swarm, a fundraising app; and proposals for the use of distributed ledgers to manage land ownership or critical infrastructures like water and energy. Many of these activities are difficult outside of local communities or in the absence of some trusted intermediary. However, I also think that much of the current rhetoric around the blockchain hints at problems with the techno-utopian ideologies that surround digital activism, and points to the assumptions these projects fall into time and again. It's worth addressing these here.

ASSUMPTION #1: WE CAN REPLACE MESSY AND TIME-CONSUMING SOCIAL PROCESSES WITH ELEGANT TECHNICAL SOLUTIONS

Fostering and scaling cooperation is really difficult. This is why we have institutions, norms, laws, and markets. We might not like them, but these mechanisms allow us to cooperate with others even when we don't know and trust them. They help us to make decisions and to divvy up tasks and to reach consensus. When we take these things away—when we break them down—it can be very difficult to cooperate. Indeed, this is one of the big problems with alternative forms of organization outside of the state and the market—those that are *not* structured by typical modes of governance such as rules, norms, or pricing. These kinds of structureless collaboration generally only work at very local kin-communal scales where everybody already knows and trusts everyone else. In Ireland, for example, there were several long-term bank strikes in the 1970s. The economy didn't grind to a halt. Instead, local publicans stepped in and extended credit to their customers; the debtors were well-known to the publicans, who were in a good position to make an assessment on their credit worthiness. Community trust replaced a trustless monetary system. This kind of local arrangement wouldn't work in a larger or more atomized community. It probably wouldn't work in today's Ireland because community ties are weaker.

Bitcoin caused excitement when it proposed a technical solution to a problem that previously required a trusted intermediary—money, or, more specifically, the problem of guaranteeing and controlling money supply and monitoring the repartition of funds on a global scale. It did this by developing a distributed database that is cryptographically verified by an entire network of peers and by linking the production of new money with the individual incentive to maintain this public repository. More recently this cryptographic database has also been used to manage laws, contracts, and property. While some of the more evolved applications involve verifying precious stones and supporting interbank loans, the proposal is that this database could also be used to support alternative worker platforms, allowing systems where people can organize, share, or sell their labor without the need of a central entity controlling activities and trimming a generous margin off the top.

Here the blockchain replaces a trusted third party such as the state or a platform with cryptographic proof. This is why hardcore libertarians and anarcho-communists both favor it. But let's be clear here—it doesn't replace *all* of the functions of an institution, just the function that allows us to trust in our interactions with others because we trust in certain judicial and bureaucratic processes. It doesn't stand in for all the slow and messy bureaucracy and debate and human processes that go into building cooperation, and it never will.

The blockchain is what we call a "trustless" architecture. It *stands in* for trust in the absence of more traditional mechanisms like social networks and co-location. It allows cooperation without trust, in other words—something that is quite different from fostering or building trust. As the founding Bitcoin document details, proof-of-work is not a new form of trust, but the abdication of trust altogether as social confidence and judgment in favor of an algorithmic regulation. With a blockchain, it maybe doesn't matter so much whether I believe in or trust my fellow peers just so long as I trust in the technical efficiency of the protocol. The claim being made is not that we can engineer greater levels of cooperation or trust in friends, institutions, or governments, but that we might dispense with social institutions altogether in favor of an elegant technical solution.

This assumption is naïve, it's true, but it also betrays a worrying politics—or rather a drive to replace politics (as debate and dispute and things that produce connection and difference) with economics. This is not just a problem with blockchain evangelism—it's a core problem with the ideology of digital activism generally. The blockchain has more in common with the neoliberal governmentality that produces platform capitalists like Amazon and Uber and state-market coalitions than any radical alternative. Seen in this light, the call for blockchains forms part of a line of informational and administrative technologies such as punch cards, electronic ledgers, and automated record keeping systems that work to administrate populations and to make politics disappear.

ASSUMPTION #2: THE TECHNICAL CAN INSTANTIATE NEW SOCIAL OR POLITICAL PROCESSES

Like a lot of peer-to-peer networks, blockchain applications conflate a technical architecture with a social or political mode of organization. We can see this kind of ideology at work when the CEO of Bitcoin Indonesia argues, "In its purest form, blockchain *is* democracy." From this perspective, what makes Uber Uber and La'Zooz La'Zooz comes down to technical differences at the level of topology and protocol. If only we can design the right technical system, in other words, the right kind of society is not too far behind.

The last decade has shown us that there is no linear-causal rela-tionship between decentralization in technical systems and egalitarian or equitable practices socially, politically, or economically. This is not only because it is technologically determinist to assume so, or because networks involve layers that exhibit contradictory affordances, but also because there's zero evidence that features such as decentralization or structurelessness continue to pose any kind of threat to capitalism. In fact, horizontality and decentralization—the very characteristics that peer production prizes so highly—have emerged as an ideal solution to many of the impasses of liberal economics.

Today, Silicon Valley appropriates so many of the ideas of the left—anarchism, mobility, and cooperation—even limited forms of welfare. This can create the sense that technical fixes like the block-chain are part of some broader shift to a post-capitalist society, when this shift has not taken place. Indeed, the blockchain applications that are really gaining traction are those developed by large banks in collaboration with tech startups—applications to build private blockchains for greater asset management or automatic credit clearing between banks, or to allow cultural industries to combat piracy in a distributed network and manage the sale and ownership of digital goods more efficiently.

While technical tools such as the blockchain might form part of a broader artillery for platform cooperativism, we also need to have a little perspective. We need to find ways to embrace not only technical solutions, but also people who have experience in community organizing and methods that foster trust, negotiate hierarchies, and embrace difference. Because there is no magic app for platform cooperativism. And there never will be.

40. NON-COOPERATIVISM

ASTRA TAYLOR

Does digital technology make the dream of a fairer, more cooperative world more possible? I'm not entirely sure that it does. To the contrary, I think there was just as good a case for cooperative ownership in the industrial age, and just as many obstacles, though perhaps those obstacles have shape-shifted. We have to take honest stock of these hurdles and confront them head on if we want to build a movement that has a chance of truly challenging the economic status quo. Of course the powerful don't want us to change things, and they will go to great lengths to stop us from doing so, whether by employing bureaucracy or brute force.

If there's one thing I want to say here it is that I want my cooperativism—platform or otherwise—to be confrontational. I think it has to be confrontational to really make a difference. Or, to put it another way, we need an inside/outside strategy: building cooperative alternatives on the margin while challenging the existing structures at the center. I'd like to see positive cooperative experiments combined with strategic campaigns of non-cooperation, of resistance to the financial system that promotes selfishness over solidarity.

By emphasizing the need not just to create alternatives, but also to confront the powers that be, I'm echoing longstanding concerns. At least since Beatrice Webb's *The Cooperative Movement in Great Britain*, published in 1891, some trade unionists have criticized cooperatives for trying to avoid the inevitable necessity of class struggle. After all, even a giant cooperative network like Mondragon has to make concessions to globalization; likewise, small, democratically run cooperatives must play by prevailing market rules on some level or close up shop. I'm sympathetic to the exigencies that lead to such compromises—we all make them in our own ways. The point, rather, is that cooperatives do not effortlessly escape the dictates of capitalism, and so they need to be part of a broader effort to challenge the dominant economic paradigm. Fostering an oppositional spirit is vital to the cooperative cause.

As we all know, we are building on old ideas here. Platform cooperativism may be new, but cooperativism isn't. Workers have dreamed of getting rid of bosses, running their own businesses, and creating a more just society since the earliest instances of labor unrest. In the 1880s, the Knights of Labor represented more than two hundred industrial cooperatives that they hoped would serve as the basis for a "cooperative commonwealth." There is also the rich history of black cooperative economic development, which is revealed by Jessica Gordon Nembhard in her excellent book *Collective Courage*, with examples ranging from the Colored Farmers' National Alliance and Cooperative Union, which had over a million members in the late 1800s, to the many efforts compiled in W. E. B. Du Bois's 1907 *Economic Co-operation Among Negro Americans*. This tradition lives on through myriad endeavors, including Mississippi's Cooperation Jackson, the Southern Reparations Loan Fund, and the Oakland & the World Enterprises, a project I'll return to.

History abounds with rousing examples of cooperative projects, and almost as many failures. What sabotaged many of these promising enterprises is lack of access to capital. Workers are more than capable of running things without the oversight of bosses or investors, but the cash or credit to purchase equipment and pay for space, machines, and materials doesn't grow on trees.

Some might say that digital technology will resolve this dilemma—you no longer need expensive heavy machinery, you only need a website—but I'm not so sure. Digital cooperatives cannot magically sidestep the economic system that provides the perverse incentives shaping the corporate online platforms many of us have problems with. Getting rid of bosses and shareholders, with their demands for short-term returns, is a big step, but only the first one. In other words, the project of creating cooperative enterprises is inseparable from creating a financial system that is productive rather than predatory, generative rather than extractive. This means, in turn, that in addition to experimenting with alternative models of business and banking, we also desperately need to transform or reform the existing economic apparatus.

The dominant paradigm of finance, in addition to putting cooperatives at a disadvantage, is what's driving much of the inequality we see today. Some insist that the problem is robots eating our jobs, but the statistics don't bear that out. The astonishing rise in inequality since the 1980s can be at least partly attributed to the explosion of salaries in the financial sector, which doubled its representation in the top 1 percent of incomes. Another big factor is pay packages for CEOs, which more than quadrupled at large companies—though presumably this won't be a problem when cooperatives take over the world.

That's assuming world conquest is on the agenda, of course. The tendency to valorize the small, local, and decentralized is something else I think we should keep debating. Decentralization is not a panacea; it does not necessarily mean distributed power or equitable distribution of wealth. This is something political theorist Wendy Brown is very insightful about in her book on neoliberalism, *Undoing the Demos*. She calls this kind of non-progressive or reactionary decentralization "devolution," and it's a term I find quite useful. "Devolved power and responsibility," she writes, "are not equivalent to thoroughgoing decentralization and local empowerment."

Which brings me to one of my final points. Centralized public options need to be on the table along with decentralized cooperative or commons-based ones. We need to think creatively about how they complement each other and how they can be combined. (Consider Janelle Orsi's proposal for a municipally owned alternative to Airbnb.) Hybrid models that connect governments and co-ops would be very much in keeping with the times, as polls show more and more people warming up to the idea of socialism. Something has shifted in a big way, and cooperative solutions that involve the state should be on the table.

But ultimately this isn't just a war of ideas; cooperativism demands we put our principles into practice. I was reminded of this when I recently spent time with Elaine Brown, who is seventy-two and ran the Black Panther Party when founding member Huey Newton was self-exiled in Cuba. In 2013 she founded the aforementioned Oakland & the World Enterprises, which will eventually be a network of cooperatively owned businesses run by and for formerly incarcerated individuals. These are the people, Brown argues, who really need to own their jobs, because they have an even harder time finding work than everyone else. Right now she has a fully functional organic farm in West Oakland, but she has plans for a cooperatively owned grocery store, fitness center, nail salon, and tech incubator under five floors of affordable housing. What does she need to get this plan to the next level? Capital, of course—though she has raised quite a bit by partnering with the city. It's a good example of the kind of municipal collaboration I think we need more of.

Elaine told me, in no uncertain terms, that one should never organize or mobilize around abstract principles. When the Panthers organized their free breakfast program, they didn't say, "You have a right to nutrition"—they fed people (and then the parents of the children they fed went and demanded that schools provide meals, because if the Panthers could do it, certainly the state could too). Likewise, she doesn't motivate her project's farmers with ideological talking points or treatises on cooperative economics; rather, she allows them to recognize how much better it is to share in the prosperity created by their labor and to be treated like true partners.

Doesn't the same hold true for everyone? Journalists, wouldn't you like to *not* be treated like disposable providers of work-for-hire content? Programmers, wouldn't you like to have a say in what you build and why you build it, and own the fruits of your labor? Professors and students, wouldn't you like to have a deeper stake in the educational institutions where you teach and learn?

Those of us who want the concept of platform cooperativism to spread should take this lesson to heart. We need to make our case by building and pointing to real examples. We also need to let go of abstractions and address concrete concerns: How will platform cooperativism make people's lives better? How will it address their real needs? How will it feed their families? Or make them feel more connected? Or maybe the real questions are: How will it make *our* lives better? How will it address *our* real needs? How will it feed *our* families? How will it make *us* feel more connected?

Asking and answering such questions will help these important ideas take root. But we have to remember to fight. Cooperation must be coupled with non-cooperation—an active resistance that complements the building of the alternatives we need.

CONTRIBUTORS

Michel Bauwens is the founder of the P2P Foundation and partner with the Commons Strategies Group. In 2014, he was research director of the FLOK Society project, which produced the first integrated Commons Transition Plan for the government of Ecuador. His recent books are *Save the World: Towards a Post Capitalist Society* with P2P (with Jean Lievens, in French and Dutch); and *Network Society and Future Scenarios for a Collaborative Economy* (with Vasilis Kostakis; Palgrave Macmillan, 2014).

Yochai Benkler is the Berkman Professor of Entrepreneurial Legal Studies at Harvard Law School, and faculty co-director of the Berkman Center for Internet and Society at Harvard University. Since the 1990s, he has played a role in characterizing the role of information commons and decentralized collaboration to innovation, information production, and freedom in the networked economy and society.

David Bollier is an author, activist, blogger, and consultant who spends a lot of time exploring the commons as a new paradigm of economics, politics, and culture. He's been on this trail for about fifteen years, working with a variety of international and domestic partners. In 2010, he co-founded the Commons Strategies Group, a consulting project that works to promote the commons internationally.

Francesca Bria is a Senior Researcher and Advisor on information and technology policy. She has a PhD on innovation economics from Imperial College, London, and an MSc on digital economy from the University of London, Birkbeck. She is the EU Coordinator of the

D-CENT project, the biggest European project on direct democracy and digital currencies. She also leads the DSI project on digital social innovation in Europe at Nesta Innovation Lab. She has been teaching in several universities in Europe, and she has advised governments, public and private organizations, and movements on technology and information policy and its political and socioeconomic impact. Francesca is an advisor for the European Commission on future internet, collective platforms, and innovation policy.

Susie Cagle is an American journalist and editorial cartoonist whose work has appeared in *The American Prospect*, AlterNet, The Awl, GOOD, and others. Cagle is based in Oakland, California.

David Carroll is an Associate Professor of Media Design at Parsons School of Design at The New School, former director of the MFA Design and Technology program, and former co-founder and CEO of Glossy, an AI brain learning culture through media. His work is situated at the intersection of art, design, media, culture, policy, science, and technology, especially in collaboration with peers in engineering, journalism, publishing, advertising, privacy, psychology, and pedagogy. He's on Twitter @profcarroll.

Miriam A. Cherry is a Professor of Law at St. Louis University and a member of the American Law Institute. Her work focuses on the intersection of technology and employment law.

Ra Criscitiello is a Research Coordinator at SEIU-UHW in Oakland, California, a union of eighty thousand healthcare workers. She is also a union-side labor attorney. She is building an innovative employment model that collectivizes the employment status of unionized workers on scale. In the new worker-driven democratic landscape, her model allows for on-demand labor without compromising traditional union values. Ra is also a surfer, triathlete, and banjo player.

John Duda is the Communications Director at The Democracy Collaborative, a national research institute developing new strategies for the democratization of wealth and the reconstruction of community in the face of systemic crisis. He lives in Baltimore, where he is a co-founder of the Red Emma's worker cooperative, and where he recently completed a PhD project examining the way the idea of self-organization developed in the sciences of cybernetics and in left politics.

Marina Gorbis is a futurist and social scientist who serves as executive director to the Institute for the Future (IFTF), a Silicon Valley nonprofit research and consulting organization. In her seventeen years with IFTF, Marina has brought a futures perspective to hundreds of organizations in business, education, government, and philanthropy to improve innovation capacity, develop strategies, and design new products and services.

Jessica Gordon Nembhard, political economist and Professor of Community Justice and Social Economic Development (Africana Studies, John Jay College CUNY), is author of *Collective Courage: A History of African American Cooperative Economic Thought and Practice* (Penn State University Press, 2014). An affiliate scholar with the Centre for the Study of Co-operatives at the University of Saskatchewan, Gordon Nembhard's memberships include GEO Newsletter, the Southern Grassroots Economies Project, and the Association of Cooperative Educators.

Karen Gregory is a Lecturer in Digital Sociology at the University of Edinburgh. Her work explores the intersection of digital labor, affect, and contemporary spirituality, with an emphasis on the role of the laboring body. Karen is a founding member of CUNY Graduate Center's Digital Labor Working Group, and her writings have appeared in *Women's Studies Quarterly*, *Women and Performance*, *Visual Studies*, *Contexts*, *The New Inquiry*, and *DIS Magazine*.

Seda Gürses is a Postdoctoral Research Associate at the Center for Information Technology at Princeton University, and an FWO fellow at the Computer Security and Industrial Cryptography (COSIC) research group at the University of Leuven in Belgium. She works on privacy and requirements engineering, privacy enhancing technologies, and surveillance. Previously, she was a postdoctoral fellow at the Media, Culture and Communications Department at the NYU Steinhardt School and at the Information Law Institute at NYU Law School. She is also a member of the feminist art collective Constant VZW in Brussels.

Steven Hill is a senior fellow at the New America Foundation and the Holtzbrinck fellow at the American Academy in Berlin. His latest book is *Raw Deal: How the "Uber Economy" and Runaway Capitalism Are Screwing American Workers* (St. Martin's, 2015), and his articles, op-eds, and interviews have appeared in *The New York Times*, *Washington Post*, *Atlantic*, *Nation*, *Wall Street Journal*, *Guardian*, BBC, *Le Monde*, Politico, CNN, C-SPAN, "Democracy Now," NPR, Salon, *Fast Company*, and more. He can be found at www.Steven-Hill.com and @ StevenHill1776.

Melissa Hoover is the founding Executive Director of the Democracy at Work Institute, which expands the promise of worker ownership to communities most affected by social and economic inequality. Prior, she served as Executive Director of the United States Federation of Worker Cooperatives. Together, this national grassroots membership organization and think-and-do tank work to build a member-based movement that reaches a scale that has real impact. Melissa worked for six years as a cooperative developer with the Arizmendi Association of Cooperatives, where she assisted in the development of two new worker-owned bakeries.

Dmytri Kleiner is a software developer and member of the Telekommunisten art group. His work investigates political economy and the social relations embedded in communications technology. He is the author of the Telekommunist Manifesto and can be followed at http://dmytri.info.

Vasilis Kostakis is a tenured Senior Research Fellow at the Ragnar Nurkse School of Innovation and Governance at Tallinn University of Technology, Estonia. He is also Founder of the interdisciplinary research hub P2P Lab and Research Coordinator of the P2P Foundation.

Brendan Martin is founder and president of The Working World. After studying economics and a stint on Wall Street, Brendan became focused on the connection of finance and economic justice. In 2004, he founded TWW in Argentina to work with the recovered factory movement for economic democracy. Brendan now heads TWW in the United States, implementing the same principles in a post-industrial context.

Michele (Micky) Metts is a member of Agaric, a worker-owned tech cooperative doing web development. She is known as an Activist Hacker/Industry Organizer/Public Speaker/Connector/Advisor and Visionary. Micky is the liaison to the Solidarity Economy Network (SEN) and the United States Federation of Worker Cooperatives (USFWC). She is a member of FSF.org and Drupal.org, a community based on free software. Micky grew up in Weston, Connecticut, and now lives in Boston with her partner, John Crisman.

Kristy Milland is community manager of TurkerNation.com. She's been a crowd worker, Requester, and researcher over the last decade. She speaks about the ethics of crowd work, exposing worker exploitation caused by a lack of legislation that would protect workers from corporate interests. She is also engaged in multiple projects to create new platforms, which will offer fair compensation and other benefits to crowd workers, and wants to join new projects which promise to do the same.

Mayo Fuster Morell directs the Dimmons Digital Commons Research Group at the Internet Interdisciplinary Institute of the Open University of Catalonia. As part of IGOPnet.cc, she is the principal investigator of the P2P Value European project on value creation in collaborative production. She is a Faculty Associate at the Berkman Klein Center for Internet and Society at Harvard University. She also directs the expertise group BarCola on collaborative economy, linked to the Barcelona City Council.

The NYC Real Estate Investment Cooperative (NYC REIC) was founded in 2015 to leverage the patient investments and political power of members to secure permanently affordable commercial properties in NYC for community, small business, and cultural use. Consistent with the principles and spirit of the cooperative movement, the NYC REIC makes long-term, stabilizing, and transformative investments for the mutual benefit of our members and our communities. Find out more at http://nycreic.org.

Rachel O'Dwyer is a research fellow and lecturer at Trinity College, Dublin. She is a member of the P2P Foundation and the IoT Council, and is the curator of Openhere, a festival and conference on the digital commons. She writes about the political economy of communications, with a focus on digital currencies and decentralized networks. She is a regular contributor to *Neural* magazine and the founding editor in chief of the open access peer-reviewed journal *Interference*. She's on Twitter @rachelodwyer.

Janelle Orsi is a lawyer, advocate, writer, and cartoonist focused on cooperatives, the sharing economy, urban agriculture, shared housing, local currencies, and grassroots finance. She is Co-founder and Executive Director of the Sustainable Economies Law Center (SELC), which facilitates the growth of sustainable and localized economies through education, research, and advocacy. Janelle is the author of *Practicing Law in the Sharing Economy* (ABA, 2014) and co-author of *The Sharing Solution* (NOLO, 2009).

Michael Peck is a co-founder of 1worker1vote.org and serves as Mondragon's North America delegate, an American Sustainable Business Council board member, a Blue Green Alliance Corporate Advisory board member, and a MAPA Group founder. 1worker1vote is a nationwide economic development catalyst, mobilizing the union-co-op hybrid model reflecting Mondragon values of single-class equity and workplace solidarity democratically practiced in profitable enterprises to overcome structural inequalities.

Carmen Rojas is the CEO of The Workers Lab, an innovation lab that invests in entrepreneurs, community organizers, and technologists to create sustainable and scalable solutions that build power for U.S. workers.

Douglas Rushkoff is an author, teacher, and documentarian who focuses on the ways people, cultures, and institutions create, share, and influence each other's values. He is Professor of Media Theory and Digital Economics at CUNY/Queens. His new book, *Throwing Rocks at the Google Bus: How Growth Became the Enemy of Prosperity*, argues that digital networks are still capable of fostering a distributed economy, but only if we can abandon the industrial-age mandate of growth above all.

Saskia Sassen (www.saskiasassen.com) is the Robert S. Lynd Professor of Sociology and Member, The Committee on Global Thought, Columbia University. She is the author of several books and the recipient of diverse awards and mentions, ranging from multiple doctor honoris causa to named lectures and various honors lists. Her new book is *Expulsions: Brutality and Complexity in the Global Economy* (Harvard University Press, 2014), which is already translated into thirteen languages, with more coming.

Nathan Schneider is a scholar-in-residence of media studies at the University of Colorado Boulder and co-organized the Platform

Cooperativism conference in 2015. His articles have appeared in publications including *VICE*, *The Nation*, *The New Republic*, *The Chronicle of Higher Education*, and *YES! Magazine*. His two books, *God in Proof* and *Thank You, Anarchy*, were published by University of California Press. Follow his work at his website, nathanschneider.info.

Trebor Scholz is a scholar-activist and Associate Professor for Culture & Media at The New School in New York City. His latest book, *Uber-Worked and Underpaid: How Workers Are Disrupting the Digital Economy* (Polity, 2016), develops an analysis of the challenges posed by digital labor and introduces the concept of platform cooperativism as a way of joining the peer-to-peer and co-op movements with online labor markets while insisting on collective ownership and democratic governance. His next book will focus on the prospects of the cooperative Internet. Follow him on Twitter at @trebors.

Juliet Schor is Professor of Sociology at Boston College and a member of the MacArthur Foundation Connected Learning Research Network. She is conducting a six-year project on economic innovations, including "sharing" spaces. A bestselling author, Schor has published books that include *The Overworked American*, *The Overspent American*, and *True Wealth: How and Why Millions of Americans Are Creating a Time-Rich, Ecologically Light, Small-Scale, High-Satisfaction Economy* (previously published as *Plenitude*).

Palak Shah is the Social Innovations Director of the National Domestic Workers Alliance, leading experimental and market-based approaches to improve working conditions, services, and employment opportunities for online workers. Palak has led NDWA's introduction into the on-demand economy with the Good Work Code, and is driving forward the organization's collaboration with Silicon Valley leaders. Previously, she served as a leader at Wellmont Health System, as a member of Massachusetts Governor Deval Patrick's administration, and as a consultant at Accenture.

Kati Sipp is the Future of Work Campaigns Director for the National Guestworkers Alliance. She created the blog *Hack the Union*, which focuses on the intersections of work, organizing, and technology. She founded the Pennsylvania affiliate of Working Families, and spent nine years working for SEIU Healthcare Pennsylvania, serving as the Political Director and Executive Vice President. Kati is the proud mother of Alina and Isaac. Follow her on Twitter at @katisipp and @ hacktheunion.

Tom Slee writes about technology and politics. His new book, *What's Yours Is Mine: Against the Sharing Economy*, was published by OR Books in November 2015. He has a PhD in theoretical chemistry, a long career in the software industry, and his earlier book *No One Makes You Shop at Wal-Mart* (BTL, 2006) is a left-wing game-theoretical investigation of individual choice that has been used in university economics, philosophy, and sociology courses.

Christoph Spehr is a politician, author, and theorist. He worked for the German National Conference on Internationalism (BUKO), the cultural co-op Paradox at Bremen, and the Left Party. From 1997 to 2005 he was co-editor of the magazine *Alaska*, and from 2008 to 2015 he was regional head of the Left Party, Bremen county. He wrote a theory of Free Cooperation that won the Rosa-Luxemburg-Prize in 2001, and was included in *The Art of Free Cooperation* (Autonomedia, 2007), along with a DVD including his video *On Rules and Monsters*.

Danny Spitzberg is a sociologist and user researcher based in Oakland, California. He is principal at peakagency.co, a collective that partners with cultural and economic justice projects to build community platforms. He is also helping build seed.coop, a platform for co-ops everywhere to grow their membership. Say hi on Twitter @daspitzberg.

Arun Sundararajan is Professor and the Robert L. and Dale Atkins Rosen Faculty Fellow at New York University's Leonard N. Stern

School of Business. He is also an affiliated faculty member at NYU's Center for Urban Science+Progress, and at NYU's Center for Data Science.

Astra Taylor is a documentary filmmaker, writer, and political organizer. She is the director of the films *Zizek!* and *Examined Life*, and the author of *The People's Platform: Taking Back Power and Culture in the Digital Age* (Picador, 2015), winner of a 2015 American Book Award. She helped launch the Rolling Jubilee debt-abolishing campaign and is a co-founder of the Debt Collective.

Cameron Tonkinwise is the Director of Design Studies and Doctoral Studies at the School of Design at Carnegie Mellon. Cameron continues to research what designers can learn from philosophies of making, material culture studies, and sociologies of technology. Cameron's research and teaching focus on the design of systems that lower societal materials intensity, primarily by decoupling use and ownership: in other words, systems of shared use.

McKenzie Wark is the author of *Molecular Red: Theory for the Anthropocene* (Verso, 2015), among other things, and teaches at The New School in New York City.

Caroline Woolard's interdisciplinary work facilitates social imagination at the intersection of art, design, and political economy. After co-founding and co-directing resource sharing networks OurGoods .org and TradeSchool.coop, Woolard is now focused on her work with BFAMFAPhD.com to raise awareness about the impact of rent, debt, and precarity on culture, and on the NYC Real Estate Investment Cooperative to create and support truly affordable commercial space in New York City.

ACKNOWLEDGMENTS

The insurgency from which this book stems, together with the book itself, could only have been possible through remarkable feats of cooperativism. Throughout this process, we have been awed by the enthusiasm, creativity, and commitment demonstrated by those who have come together to help make a more cooperative online economy a reality. The book's contributors, who offered their insights in the essays and their ingenuity in the showcases, have been inspiring and patient collaborators. Many other people, too, have helped support this effort, often in less visible ways.

Micah Sifry was an early adopter of Trebor Scholz's concept of platform cooperativism, and he, together with his colleague at Civic Hall, Andrew Rasiej, were a source of ongoing support, advice, and meeting space. Neal Gorenflo of Shareable was among the first to recognize that a real, cooperative sharing economy was starting to take shape; he helped guide Nathan Schneider into discovering it for himself. Palak Shah of the National Domestic Workers Alliance and Sara Horowitz of the Freelancers Union were among our early conversation partners who helped connect this work to the challenges workers are facing at the front lines of the gig economy, as did two courageous worker-organizers on Amazon's Mechanical Turk, Rochelle LaPlante and Kristy Milland. Natalie Foster of Institute for the Future provided invaluable guidance as well.

We are grateful for the support of this unconventional research project from colleagues at our institutions, especially Dean Stephanie Browner, Karen Noyes, Alexander Draifinger, and Verna De LaMothe at The New School's Lang College and, at the University of Colorado Boulder's College of Media, Communication, and Information, Dean Lori Bergen and Nabil Echchaibi.

The book itself came about thanks to the willingness of John Oakes at OR Books to say yes and ask questions later, as well as his commitment to the challenge of creating righteous models for independent publishing in the age of Amazon. And it is only thanks to the rigor and guidance of Samira Rajabi that we were able to meet a seemingly impossible self-imposed deadline.

The Platform Cooperativism event in November 2015 was sponsored by Eugene Lang College The New School for Liberal Arts, Demand Progress, the Democracy at Work Institute, the Ford Foundation, the Freelancers Union, IG Metall, the Institute for the Future, Internet Society, the Lang Student Senate, The New School's Digital Humanities Minor, The New School University Student Senate, the Robert L. Heilbronner Center for Capitalism Studies, University of Colorado Boulder's College of Media, Communication, and Information, the Rosa Luxemburg Foundation NYC, and the Workers Lab.

The event was presented in partnership with Carnegie Mellon School of Design, Civic Hall, Democracy Collaborative, Green Worker Cooperatives, The Laura Flanders Show, the Murphy Institute for Worker Education and Labor Studies at CUNY, the New Economy Coalition, OccupyWallStNYC, the Robin Hood Foundation, Shareable, the United States Federation of Worker Cooperatives, Ver.di, The Working World, and the Yale Information Society Project.

With far more attendees than originally planned for, we also relied on the work of intrepid volunteers, including Lauren Mobertz, Meg Miner, Angela Difede, and countless others.

Throughout this process we have relied on the collaboration, guidance, and support of our families—Claire and Daniel Francis, and Jenny, Rosa Clara, and Emma Luisa. They kindle our hope in a more cooperative world and, for us, prove that it is possible.

FURTHER RESOURCES

LAUNCH EVENT

"Platform Cooperativism: The Internet, Ownership, Democracy," The New School (November 2015), video archive: http://platformcoo p.net/2015/video

READINGS

Trebor Scholz, "Platform Cooperativism vs. the Sharing Economy," Medium (December 5, 2014), https://tinyurl.com/oj8rna2

Nathan Schneider, "Owning Is the New Sharing," *Shareable* (December 21, 2014), http://shareable.net/blog/owning-is-the-new-sharing

Janelle Orsi, Frank Pasquale, Nathan Schneider, Pia Mancini, Trebor Scholz, "5 Ways to Take Back Tech," *The Nation* (May 27, 2015), http://thenation.com/article/5-ways-take-back-tech

Trebor Scholz, *Platform Cooperativism: Challenging the Corporate Sharing Economy* (Rosa Luxemburg Stiftung, New York Office, 2016, with additional translations in Spanish, French, Portuguese, German, Italian, and Chinese), http://platformcoop.net/about/primer

Trebor Scholz, *Uberworked and Underpaid: How Workers are Disrupting the Digital Economy* (Polity, 2016)

WEBSITES

Platform Cooperativism portal, http://platformcoop.net

Platform Cooperativism Consortium, http://platformcoop.newschool.edu

The Internet of Ownership, http://internetofownership.net

Shareable, http://shareable.net

Sustainable Economies Law Center, http://theselc.org